LEVELS OF INFINITY

SELECTED WRITINGS ON MATHEMATICS AND PHILOSOPHY

HERMANN WEYL

TRANSLATED AND EDITED WITH AN
INTRODUCTION AND NOTES BY
PETER PESIC
St. Johns' College
Santa Fe, New Mexico

DOVER PUBLICATIONS, INC.
Mineola, New York

Bibliographical Note

A compilation of essays by Hermann Weyl, *Levels of Infinity* is a new work, first published by Dover Publications, Inc., in 2012.

International Standard Book Number

ISBN-13: 978-0-486-48903-2
ISBN-10: 0-486-48903-5

Manufactured in the United States by LSC Communications
48903504   2023
www.doverpublications.com

Contents

Contents

Introduction

"A Proteus who transforms himself ceaselessly in order to elude the grip of his adversary, not becoming himself again until after the final victory." Thus Hermann Weyl (1885–1955) appeared to his eminent younger colleagues Claude Chevalley and André Weil. Surprising words to describe a mathematician, but apt for the amazing variety of shapes and forms in which Weyl's extraordinary abilities revealed themselves, for "among all the mathematicians who began their working life in the twentieth century, Herman Weyl was the one who made major contributions in the greatest number of different fields. He alone could stand comparison with the last great universal mathematicians of the nineteenth century, Hilbert and Poincaré," in the view of Freeman Dyson. "He was indeed not only a great mathematician but a great mathematical writer," wrote another colleague.[1] This anthology presents a spectrum of Weyl's later mathematical writings, which together give a portrait of the man and the mathematician. The works included have been chosen for their accessibility, but they do also include, where needed, the mathematical details that give vivid specificity to his account. Weyl's essays will convey pleasure and profit both to general readers, curious to learn directly from a master mathematician, as well as to those more versed, who want to study his unique vision. Those who wish to explore further his seminal work in physics in its philosophic and mathematical contexts will find relevant writings in a companion anthology, *Mind and Nature* [2009a], which gathers his writings in these fields; as a result of Weyl's inclusive vision, these two anthologies overlap and complement each other at many points.

The protean Weyl drew his many shapes from rich and complex life-experience, in which mathematics formed only one strand in a complex tapestry. Already philosophic in temperament as a teenager who pored over Kant, Weyl was deeply formed by his work with David Hilbert, his teacher and mentor at the University of Göttingen. Weyl recalled being "a country lad of eighteen" who became entranced by Hilbert, a Pied Piper "seducing so many rats to follow him into the deep river of mathematics."[2] Widely considered to have been Hilbert's favorite student, Weyl graduated in 1910 and stayed on as a *Privatdozent* until 1913, when he took a post at the Eidgenössische Technische Hochschule (ETH) at Zürich.[3]

There he met Albert Einstein, who during 1913–1914 was in the midst of

1

his struggle to generalize relativity theory, which led him to study Bernhard Riemann's sophisticated mathematics of curved higher-dimensional manifolds. Returning to Switzerland in 1916 after his military service, Weyl's "mathematical mind was as blank as any veteran's," but Einstein's general relativity paper of that year "set me afire." Weyl went on to write his "symphonic masterpiece" (as Einstein called it), *Space-Time-Matter* [1918b], one of the first and perhaps still the greatest exposition of relativity. As part of his explanatory work, Weyl went on to formulate his seminal gauge theory unifying electromagnetism and gravitation [1918c], which Einstein hailed as "*a first-class stroke of genius.*"[4] Though the initial hopes for this theory faded in the light of difficulties noted by Einstein (and later the need to incorporate quantum theory), Weyl continued to ponder the generalizations and implications of his 1918 theory. Looking back, Weyl's work laid the foundation of the gauge theories that, fifty years later, unified the strong, weak, and electromagnetic theories, fulfilling his initial aspiration in ways he had not dreamed.

Weyl eventually succeeded Hilbert at Göttingen, though only after protracted hesitation: first offered a professorship there in 1923 (to replace Felix Klein, another great mathematician important to his development), Weyl refused, happier to remain in Zürich, where he could find relief for his asthma in the mountains. He finally accepted the call to Göttingen in 1930, when Hilbert himself retired. The first essay in this volume dates from that year, as the newly-appointed Weyl travelled to Jena to speak to mathematics students. This essay, "Levels of Infinity," was not included in Weyl's collected papers and has not been reprinted since 1931 nor ever before translated into English.[5] It gives an arresting portrait of his intellectual development and his struggle with basic questions in the foundations of mathematics.

Though devoted to Hilbert, Weyl for a time sided against him in an ongoing controversy that continues to simmer under the surface of mathematical practice. In the course of introducing unheard-of levels and degrees of infinity during the 1880s, Georg Cantor opened new mathematical realms that excited but also disturbed thoughtful practitioners.[6] Against Cantor, his eminent contemporary Leopold Kronecker argued that only the integers had fully authentic mathematical reality, challenging innovators to express their new insights in terms of this mathematical bedrock or else abandon them as unsound, following his motto "God created the integers; everything else is the work of man."[7] L.E.J. Brouwer, a distinguished topologist four years older than Weyl, made this cause his own, under the banner of "intuitionism": mathematics should return to its roots in human intuition, radically restricting its use of infinite quantities to those that could be grounded in possible acts of intuition. This would mean the exclusion of speculatively constituted sets, such as the set of all sets of real numbers, for which no intuition could possibly exist.[8]

To that end, Brouwer disallowed the use of the "principle of the excluded middle"— that any proposition either is true or its negation is true — especially when applied to infinite sets. This principle could be applied to argue that, if the negation of some mathematical proposition is demonstrably false, therefore the proposition must be true merely by elimination, without providing a positive demonstration. Brouwer advocated abandoning such propositions as not proven. For instance, let us examine the proposition "there exist two irrational quantities a and b such that a^b is rational." The Pythagoreans had already shown that $\sqrt{2}$ is irrational. Consider the quantity $\sqrt{2}^{\sqrt{2}}$; by the principle of the excluded middle, it is either rational or irrational. If it is rational, the proof is complete, for it exemplifies the quantity a^b sought in the case $a = b = \sqrt{2}$. If, on the contrary, $\sqrt{2}^{\sqrt{2}}$ is irrational, then set $a = \sqrt{2}^{\sqrt{2}}, b = \sqrt{2}$, so that

$$a^b = (\sqrt{2}^{\sqrt{2}})^{\sqrt{2}} = \sqrt{2}^{\sqrt{2} \cdot \sqrt{2}} = \sqrt{2}^2 = 2,$$

which is rational, thus implying that the proposition is true. From Brouwer's point of view, though, the assumption that $\sqrt{2}^{\sqrt{2}}$ is rational must be supported by positive proof, not just used as a hypothesis in a proof by exclusion.[9] Merely side-stepping the issue by moving the symbols around does not address the underlying question:*what sort of existence* does $\sqrt{2}^{\sqrt{2}}$ have?

This critique had deep implications for the whole conduct of mathematics. For instance, Cantor had conjectured that no cardinal number lies between those denoting the countability of the integers, $\aleph_0$, and the uncountable continuum of real numbers, $\mathfrak{c} = \aleph_1$. Brouwer considered this "continuum hypothesis" meaningless because the idea of "numbering" the continuum is an empty, abstract construct that lacks any intuitive substance, compared to our intuition of the integers. Most of set theory and large parts of analysis would have to be vacated, on Brouwer's view. Ludwig Wittgenstein was also among those who considered that set theory was "pernicious" and "wrong" because it leads to "utter nonsense" in mathematics.[10]

Weyl was deeply struck by Brouwer's arguments and addressed the problematic groundwork of mathematics in his seminal book *The Continuum* (1918), avoiding the most problematic recourse to nondenumerable infinities through a redefinition of the concept of number in terms of denumerable Cauchy sequences; Weyl will return to this idea in several of the papers in this volume. Weyl's work on the mathematical continuum is notably simultaneous with his daring attempt to generalize the space-time continuum through his invention of gauge theories; a preoccupation with seamless continuity versus atomism runs through both streams of his thought.[11]

In the early 1920s, Weyl went so far as to set his own foundational approach aside and acclaim Brouwer as *"die Revolution."*[12] Hilbert, in contrast, advocated a purely abstract formalism in which mathematics became a meaningless game played with symbols, utterly detached from intuition and hence untainted by human fallacies and illusions. Hilbert thought thereby to assure at least the non-contradictoriness of mathematics, leaving for the future to prove its consistency through iron-clad logical means that would owe nothing to mere intuition.[13] For him, Weyl's metamorphosis into an acolyte of intuitionism verged on betrayal of what Hilbert thought was the essential mathematical project, which included the "paradise" (as Hilbert called it) of Cantor's transfinite numbers. But by the mid-1920s, Weyl's initial enthusiasm for intuitionism had given way to a more measured view of "the revolution," which he (along with Hilbert) judged would leave in ruins too much beautiful and important mathematics that could not be proved using intuitionistically pure arguments.

"Levels of Infinity" gives an important portrait of Weyl's views in the wake of this critical period. Now speaking as Hilbert's successor at Göttingen, Weyl aims to give a larger, more inclusive view that will do justice both to the core of intuitionistic insight, in which he still believes, and yet take into account Hilbert's formalism. From the first line, Weyl embraces mathematics as "the science of the infinite" but at the same time emphasizes his sense of "the impossibility of grasping the continuum as a fixed being." Weyl sets forth a series of levels of infinity, giving "an ordered manifold of possibilities producible according to a fixed procedure and open to infinity," an openness he eloquently celebrates as the essence of mathematical — and human — freedom: "Mathematics is not the stiff and paralyzing schema the layman prefers to imagine; rather, with mathematics we stand precisely at that intersection of bondage and freedom that is the essence of the human itself."[14]

His critique of the actual, finished infinite leads Weyl to advocate explicit *construction* of mathematical entities as the antidote to the "mathematically empty" strategem of considering existent whatever is not categorically impossible: "Only by means of this constructive turn is a mathematical mastery, an analysis, of continuity possible." Weyl's philosophic seriousness cannot brook the triviality he finds in Hilbert's depiction of mathematics as a meaningless game with empty symbols. Instead, Weyl emphasizes the profound implications of the symbolic mode of constructive mathematics, which he connects with the pervasive philosophic implications of symbolic forms throughout human thought and expression.[15] Though strongly influenced by his contact with the phenomenological philosopher Edmund Husserl, Weyl's sense of the priority of symbolic expression goes beyond what is explicable merely through the phenomenal flux of experience: "We cannot deny that there lives in us a need for theory that is absolutely inexplicable from a merely phenomenalist standpoint. That need has a drive to create directed to a symbolic shaping of the transcendent, which demands satisfaction .... All creative shaping by humans receives its deep holiness and dignity from this connection." Though he disclaimed theism as well as atheism, Weyl read attentively the writings of Meister Eckhart and thought deeply about the connection between theology and mathematics, as when he describes "God as the completed infinite, [which] can neither break into the human through revelation nor can the human by mystic contemplation break through to Him. We can only represent the completed infinite in the symbol."[16]

In 1931, Weyl spoke on "Topology and Abstract Algebra as Two Roads of

Mathematical Comprehension" at a summer course for Swiss teachers, showing his interest in improving secondary instruction and guiding its development. He begins by invoking his teacher Felix Klein, who had been much involved in this cause.[17] The problem of intuition is foremost in Weyl's mind; he notes that we do not merely want to follow formal reasonings blindly but "want to understand the *idea* of the proof, the deeper context," not just its "machinery." Weyl praises Klein's "intuitive perception" and brings forward topology as exemplifying this kind of visually compelling approach to mathematics. At the same time, though, Weyl notes how the algebraic path, though less intuitively appealing, completes and makes rigorous what topology had discerned more generally. His image of two *Wege*, paths or roads, evokes the sense of their different, even contradictory, qualities, though in many of his examples he wishes to stress that these two ways ultimately complement each other. His conclusion, though, sides with topology as providing the essential initial insight into what he calls the "mathematical substance" that only afterwards is fruitfully mined by algebraic labor. The image of mining also lies behind his rather dark ending, which speaks of the exhaustion of the abstract vein and the hard times he foresees for mathematicians.[18]

In 1939, he phrased this alternative much more starkly: "In these days the angel of topology and the devil of abstract algebra fight for the soul of each individual mathematical domain." Those were not the only angels and devils struggling at the time. The intervening dark years included the tragic ending of the great period of Göttingen mathematics, which Weyl experienced first-hand. He had returned there reluctantly, having been happy since 1913 in Zürich, where he felt greater freedom than he had in Germany. His hesitation was prescient; the growing darkness of the early 1930s extinguished the brilliance of Göttingen, whose Jewish mathematicians were systematically stripped of their offices and attacked by the Nazi racial laws. By 1933, Weyl, with his Jewish wife and liberal sympathies, left for the newly-founded Institute for Advanced Studies at Princeton, where his colleagues included Einstein, Kurt Gödel, and John von Neumann.[19]

This dark history shadows Weyl's portrait of Emmy Noether (1882–1935), his contemporary, friend, and mathematical colleague, who also had fled to the United States, though she only lived two years after taking up a position at Bryn Mawr. Delivered as a memorial address after her early death, Weyl brings to life an extraordinary figure he praises as mathematician who struggled against prejudice, yet without becoming embittered. From our present distance in time, we may be disturbed by the categorization of her as a "woman mathematician" implicit in Albert Einstein's description of her as "the most significant creative mathematical genius thus far produced since the higher education of women began" or Weyl's description of Noether as "a great mathematician, the greatest, I firmly believe, her sex has ever produced, and a great woman."[20] Rather than being condescending, Einstein and Weyl were trying to do justice to her situation, and that of women scientists in general, in light of the historical context and the prevalent prejudices against them. Hilbert himself had stood up for her: "I do not see that the sex of the candidate is an argument against her admission as *Privat-dozent*. After all, the [Göttingen Academic] Senate is not a bath-house."[21] But even Hilbert's strong advocacy only resulted in an irregular, unpaid appointment for her.

Though Weyl shows the injustices she endured, his eulogy gives a detailed account of her mathematical accomplishments, as well as her personality and unique life-trajectory. Here, too, he takes on the mantle of Hilbert as Noether's champion, now laboring not to win her the kind of university post to which her merits entitled her but to give her the posthumous recognition she now enjoys. Thanks to Noether, and to Hilbert, Einstein, and Weyl, now those passionate about science and mathematics can pursue it in light of her example, the heritage of prejudice increasingly relegated to the all-too-human past.

Weyl's account of Noether situates her in the vortex of the eventful history of modern mathematics, her insights forming part of complex developments that go far beyond the "merely-personal" aspects of gender, religion, or nationality. Physicists may know best Noether's principle connecting the symmetries of dynamics with conserved quantities (as symmetry in time-displacement is manifest in conservation of energy, for instance), which Weyl embeds in a much larger picture of her work. His overview will help contemporary readers gain a fuller picture of her accomplishments, which have deep implications for mathematics and physics both. Weyl emphasizes her wide range, which included heroic formal computations and conceptual axiomatic thinking; as such, she exemplifies his recurrent theme of the complementary interdependence of axiomatic and constructive approaches to mathematics. "She originated above all a new and epoch-making style of thinking in algebra," the generalized modern approach, which Weyl above contrasted with the intuitive sweep of topology. An influential teacher, Noether's students and followers carried her influence into the ensuing developments in mathematics and its increasingly "abstract" style.[22]

Weyl's eulogy of Noether was written in English, as were the remainder of the works in this anthology, befitting his new life in the United States. Here and at many other points of his subsequent English writings, Weyl revels in the idioms and possibilities of his new tongue; while still remaining very characteristic of his personal style, his English writings are notably clearer than his often intricate, difficult German prose. His ability to adapt to a new land and a new language, indeed to revel in his rejuvenation and new life, were not the least of his protean metamorphoses.

In his new American phase, Weyl became a significant figure in public intellectual life, though never very well-known or widely popular; he neither received nor craved the *réclame* that dogged Einstein. He gave celebratory addresses on a number of occasions, including the bicentennial of the University of Pennsylvania (1940), for which he delivered "The Mathematical Way of Thinking."[23] Weyl emphasizes the larger significance of mathematics for the sciences and everyday thinking, even above its internal workings, left to itself. Where Brouwer was cool to applied mathematics, Weyl had long been deeply interested in physics and the broadest implications of mathematics. Consistent with his own practice, he refuses to divide truth "into watertight compartments like historic, philosophical, mathematical thinking, etc. We mathematicians are no Ku Klux Klan with a secret ritual of thinking," using a vivid allusion to his new American milieu to make his larger point. Mathematics, like democracy, is a "way of life," not a lifeless set of formal axioms nor purely introspective intuitions, à la Brouwer.[24]

Weyl critiques the common description of mathematics as "abstract," breathing thin air and reducing everything to thinner outlines. On the contrary, mathematics requires the man in the street "to look things much more squarely in the face; his belief in words must be shattered; he must learn to think more concretely." The full potency of "purely symbolic construction" lies behind what we unthinkingly call "abstraction." Weyl's charming example of the exact height of Longs Peak shows that the "abstract" concept of potential is, in fact, more concrete than the common notion of altitude. Mindful of "the witchcraft of words in the political sphere," Weyl argues that "the scientist must thrust through the fog of abstract words to reach to concrete rock of reality." To do so, "the intuitive picture must be exchanged for a symbolic construction," reminding us also of Weyl's own reconsiderations of the role of intuition in mathematics, by comparison with the power of symbols.[25]

His ensuing examples include relativity theory, basic concepts of number, and topology, whose natural connections and analogies are foremost in his mind, rather than any rigid separation between "pure" and "applied" mathematics. Looking back to his 1918 work on the continuum, one can here see that he continues to think about ways in which an "atomic" view might illuminate the troubled foundations of mathematics. He looks at sphere and torus with the eyes of a physicist, even as he contemplates space-time with a mathematician's gaze. As he quotes Galileo, one recognizes his kinship with that archetypal mathematician-philosopher. Weyl recognizes in the basic mathematical concept of *isomorphism* the essence of what physicists call *relativity*. Symbols can encompass and unify both concepts: here the axiomatic and the constructive aspects of mathematics converge.

Hilbert remained behind in Germany, retired unhappily in the desolate ruin of Göttingen, in his "tragic years of ever deepening loneliness" after 1933. Though he had been deeply disturbed by Weyl's temporary adherence to "the Revolution," Hilbert continued to esteem his "favorite son," especially after Weyl drew away from Brouwer's intransigence. For his own part, Weyl said that his motto remained: "True to the spirit of Hilbert." On the occasion of Hilbert's seventieth birthday (1932), Weyl wrote: "Woe to the youth that fails to be touched to the core by such a man as Hilbert!"[26] Hilbert died in 1944, in the depths of the war. Weyl memorialized him in two essays, the first less detailed and more personal, the second a deep and wide-ranging survey that covers, in broad outline (but with considerable richness of detail), Hilbert's preoccupations and achievements.

These are complementary accounts and can be read in different ways, according to the reader's desire to penetrate more or less deeply into the substance of Hilbert's work. Both are suffused with Weyl's own distinctive style, eloquent, profound, and more expansive than we are used to in our hurried times. Even when Weyl is describing the content of a mathematical theory, he writes personally and expressively; he *feels* the direction and import of mathematical ideas no less than the force of human character. Indeed, for Weyl the mathematical and the human converge not merely in the context of anecdote or recollection but in his keen response to the felt impact of both. Weyl registers surprise and wonder, both at human and mathematical beings. As he notes about the deep problems of the foundations of mathematics, "'mathematizing' may well be a creative activity of man, like language or music, of primary originality, whose historical decisions defy complete objective rationalization."[27]

Weyl's portrait of Hilbert is filled with amazing insights, from his close perspective as student and colleague over many years. Who else could have described "the peculiarly Hilbertian brand of mathematical thinking" as "a swift walk through a sunny open landscape; you look freely around, demarcation lines and connecting roads are pointed out to you, before you must brace yourself to climb the hill; then the path goes straight up, no ambling around, no detours." The words Weyl uses to describe how Hilbert transformed geometry eloquently describes his own amazing accomplishment in this essay: through his consummate artistry and insight, we see Weyl, no less than Hilbert, "as if one looked into a face thoroughly familiar and yet sublimely transfigured." This comes about through Weyl's familiarity with every side of Hilbert's work — indeed, with every side of mathematics — which he seems to have lived through and *felt*, not just mastered.[28]

Weyl's amazing use of words deserves special attention, and not merely because of the great beauty of what he says. He understands that, even in the seemingly trans-linguistic realm of mathematics, words are essential tools, especially in light of Hilbert's radical reconsideration of its formalism as a kind of game. As Weyl notes, "already in communicating the rules of the game we must count on understanding. The game is played in silence, but the rules must be *told* and any reasoning about it, in particular about its consistency, communicated by *words*."[29] In his discussion of Hilbert's new axiomatics, Weyl thus deepens the connection between his own symbolic view — within which words play a central role — and Hilbert's formalism.

Weyl also emphasizes the breadth of Hilbert's accomplishment; though Hilbert is too consistent and forthright to be a shape-shifter, the many-sided Weyl is specially equipped to speak about his teacher's manifold contributions across mathematics. In each section, Weyl himself seems transformed into the ideal guide in each different section, precise and incisive in his discussions of algebra, profoundly discursive about foundational issues, steeped in Fredholm's theory when integral equations come to the fore. Even the preoccupation with physics is shared between Hilbert and Weyl, showing another aspect of their temperamental similarity.

"A tree is best measured when it is down," the saying goes, but when in 1946a Weyl wrote his preface to a review [1946b] of a volume of essays about Bertrand Russell, his subject was still very much alive and active as a public figure. Weyl takes this occasion to give a wonderful overview of the realms of modern logic, to which Russell had contributed so signally.[30] Russell's own lucidity is equalled by the clarity with which Weyl recounts his ideas, including touches of wry humor (such as comparing Russell's theory of types to the dogmas of the Church Fathers, as little as Russell could be accused of religious dogmatism or indeed of any kind of conventional religiosity). Such humor aside, the cosmopolitan Weyl acclaims his no-less-worldly friend for the "incomparable riches" of U, the "Russell universe." Russell's scheme of setting up a hierarchy of types in order to avoid antinomies and paradoxes (such as Russell himself had pointed out) might be compared with the levels of infinity that Weyl himself had considered. Still, Weyl is quite clear about what he considers the limitations of the project of Russell and Whitehead's *Principia Mathematica*. Throughout the essay, Weyl seeks to find ways to reconcile or at least confront the axiomatic approach Russell pioneered with the constructive approach Weyl shared with the intuitionists. Weyl ends with a half-humorous diagram that locates his own "universe" W on a diagram, quite a bit to the left of Russell's U and Ernst Zermelo's Z, and hovering above Brouwer's B and Hilbert's H.[31]

On March 19, 1949, Weyl joined in the celebration of Einstein's seventieth birthday; a photograph of that occasion shows a somewhat bemused Einstein surrounded by colleagues, Weyl and Gödel standing close by.[32] Weyl's lecture on "Relativity Theory as a Stimulus in Mathematical Research" reflects his long and deep involvement with Einstein's ideas, which "made an epoch in my own scientific life." Weyl emphasizes the long mathematical search for *invariants*, of which relativity was one facet, along with general geometric invariants and (more surprisingly) even the Galois theory of algebraic solvability. Indeed, Einstein himself had come to realize that the name "invariant theory" would have been a far more accurate choice than "relativity theory," with its unfortunate associations of indiscriminate relativism and "anything goes." Weyl gives considerable attention to the work of Élie Cartan, with whom Einstein had an extended correspondence and who, along with Weyl, had demonstrated the naturalness of general relativity as a generalization of Newtonian mechanics. With characteristic modesty and irony Weyl describes himself as "a lone wolf in Zurich" who "was all too prone to mix up his mathematics with physical and philosophical speculations."[33]

In 1951, Weyl wrote his magisterial survey of "A Half-Century of Mathematics," an overview for which he was uniquely qualified, having participated in so many different facets of its development during the five decades of his own mathematical life. In a special way, this is a kind of intellectual autobiography, as if Proteus were to reprise his multiform career, blending his many shapes with the larger stream of mathematical imagination in which he had been so deeply immersed, which "has the inhuman quality of starlight, brilliant and sharp, but cold." Weyl looks back to the beginnings of his work (and of our anthology) when he begins with mathematics as "the science of the infinite," but now, decades later, his narration has acquired many more layers of meaning and levels of nuance. For instance, he emphasizes that certain concepts like set and mapping are really *premathematical*, even though they often appear in texts as if they were simply part of the fabric of mathematics itself. There are also practical, witty touches: he explains the order of inverse transformations $T^{-1}S^{-1}$ by comparison to the order in which one dresses and undresses, ST, where $S =$ shirt and $T =$ jacket (in that more formally attired day). Weyl constantly connects his exposition with references to the physics of music, light, thermodynamics, and quantum theory. Even in the midst of his discussion of "pure" mathematical fundamentals, he notes that "the general problem of relativity consists in nothing else but to find the group of automorphisms." He vividly compares the "wildness" or "tameness" of different groups and their representations (a field to which he himself made signal contributions that he, characteristically, does not mention). Without mentioning his own name, Weyl also seems to bid farewell to his own dreams of a unified field theory: "it is probably unsound to try to 'geometrize' all physical entities."[34]

Weyl returned to many of these themes and placed them in an even larger context in his 1953 essay on "Axiomatic Versus Constructive Procedures in Mathematics," here published for the first time in its complete, original form. He is especially troubled that "our mathematics of the last decades has wallowed in generalizations and formalizations," which he explains as a search for simplicity. Despite the general tendency to dilute the "good nourishing soup" of mathematics with "cheap generalizations," Weyl still praises "the basic soundness and importance of the axiomatic approach." In this essay, he brings into the foreground a theme that had run through many of his earlier writings, the ultimate complementarity of the axiomatic and constructive approaches. Though the desire to do justice to axiomatics underlies his even-handed presentation, ultimately he confesses his own long-standing predilection, that "my own heart draws me to the side of contructivism."[35]

This anthology ends with what may well be Weyl's last essay, which has never before appeared in print. He delivered "Why is the World Four-Dimensional?" as a lecture in Washington, D.C. on March 29, 1955, only nine months before his death of a sudden heart attack soon after his seventieth birthday.[36] After retiring from the Institute for Advanced Study in 1951, Weyl had been going back and forth between America and Switzerland. Returning to the site of his youthful freedom, Weyl revisited some of the thoughts that had germinated there. During the course of his intensive work on general relativity in 1918, he had begun to think about what deeper mathematical and physical reasons might underlie the four-dimensionality of space-time. In the course of introducing his arguments about gauge invariance, he also confronted the question of conformal invariance, which arbitrarily rescales the fundamental unit of length. Weyl was also versed in the conformal mappings of complex variables, which he explored in his pioneering book *The Idea of a Riemann Surface* (1913); here again his protean perspective enabled him to make connections between seemingly diverse fields. From these considerations, in 1919 Weyl had suggested that the mathematical structure of Maxwell's equations necessitated three dimensions of space and one of time as related in the four-dimensional manifold introduced by his teacher Hermann Minkowski (1908). As usual, Weyl presents his conclusion with unforgettable style:

> Suppose that a single light, a candle, is burning in the world. Now blow this candle out; what will happen according to the Maxwellian laws? You probably think it will grow dark, pitch dark, in a sphere around the candle which expands with the velocity of light. And you are right — *provided the number n is even, especially in our world for which n = 4.* But it would not be so in a world of odd dimensionality. Now here you have a physically interesting difference at least between even and odd dimensions, although it does not single out the dimensionality 4. "And God said, Let there be light: and there was light," so tells the story of Creation in Genesis, Chapter I. If He wished to keep the possibility open for Himself to say "Let there be darkness again" and to accomplish this by blowing out His candles then He had to make the world of even dimensionality.[37]

Weyl's arguments continue a long stream of preceding speculations about dimensionality, reaching back to Aristotle, Galileo, and Kant.[38] In this 1955 essay, Weyl reprised his youthful insights, noting the limitations of his early arguments based on Maxwell's equations but also adding new and suggestive topological arguments. In the twenty-first century, these issues have become increasingly important in the context of string and other theories using higher-dimensional manifolds that need to be connected with four-dimensional space-time. For all these initiatives and for the burgeoning study of conformal field theories, Weyl was, yet again, the pioneer who first gazed on the promised land.[39]

Aside from the two essays translated from German, the other texts follow Weyl's printed English version or manuscript without change in punctuation; though his practice does not follow presently standard usage, it seemed better to allow the reader access to his original text. Foreign language terms have been italicized according to current practice, but otherwise Weyl's original use of italics has been preserved. Notes follow each essay, clarifying the sources of the texts and offering some comments and references, including Weyl's own. I am responsible for all material in square brackets [···]; though Weyl's original citations have been kept as he had them, other editorial references are keyed to those given in full at the end of the book. I thank John Grafton for his interest and support of this and so many other books we have done together. I am deeply grateful to Brandon Fogel, Jeremy Gray, Paolo Mancosu, Erhard Scholtz, Thomas Ryckman, and Norman Sieroka for their friendly advice; Philip Bartok, Brandon Fogel, and Norman Sieroka were most generous and helpful in revising the German translations, as was Abe Shenitzer who kindly allowed the reprint here of his translation of one of the essays. This book owes a great deal to the superb technical and editorial collaboration of Alexei Pesic, whom I particularly thank. Finally, I sincerely thank Hermann Weyl's grandchildren Annemarie Weyl Carr, Peter Weyl, and Thomas Weyl for their gracious interest and their permission to publish these works. Weyl's sons Michael and Joachim, each extraordinary in his own right, have passed on; I remember with gratitude lively conversations with Michael and dedicate this anthology to their memories.

Raoul Bott recalled that "Weyl's seminar lectures were not particularly easy to understand, but they always had about them this air of ease, of a natural motion in the inevitable stream of the subject itself.... At a crucial point of a lecture, Hermann Weyl had the habit of lifting his shoulders and then letting them fall again. This motion seemed to convey the inevitability of that particular turn of thought, the God-given nature of our subject, and the minor role that he himself might have had in its development." Looking back on the whole arc of Weyl's career, Dyson judged that "so long as he was alive, he embodied a living contact between the main lines of advance in pure mathematics and in theoretical physics. Now he is dead, the contact is broken, and our hopes of comprehending the physical universe by a direct use of creative mathematical imagination are for the time being ended."[40] May this collection help bring Weyl's protean imagination back to life for a new generation.

PETER PESIC

Notes

[1] Chevalley and Weil 1957; *WGA* 4:655; Dyson 1956, 457; Newman 1957, 308.

[2] See below, 95

[3] See Weyl 1953 for a valuable account of the German university system especially as he had known it before the Second World War: "the *Privatdozent*, unlike the professor, is ... not an appointee of the state.... He has the right to lecture but no obligations whatsoever. Therefore, he can devote his whole time and energy to research and to giving lectures ... on such topics as are of interest to him." For Weyl and the ETH, see Frei and Stammbach 1992.

[4] Weyl's recollections come from 2009a, 168; the Einstein quotes from Einstein 1987–, 8:669–670 [491]; 8:710 [522] (emphasis original). Regarding the discussions between Einstein and Weyl about his 1918 theory, see Ehlers 1988, Scholz 1994,

1995, 1999a, 2001b, 2004, 2009 Sigursson 1991, Fogel 2008. On the history of gauge theory, see Scholz 1980, 1994, 1999a, 1999b, 2001b, 2004, O'Raifeartaigh 1997, 2000, Ryckman 1996, 2003, 2007, Straumann 2001.

[5] It should be noted that there is considerable overlap between this essay and Weyl's lecture on "Infinity" from *The Open World*, Weyl 2009a, 66–82.

[6] Ewald 1996, 2:838–940 contains an excellent selection of Cantor's writings and correspondence.

[7] For Kronecker, see Ewald 1996, 2:941–955, which discusses the source of his famous aphorism on 942; see also Weyl's comment on 94, below, and Edwards 1995.

[8] For helpful overviews of the intuitionistic controversy (and Weyl's place in it), see Kleene 1967, 191–201; van Dalen 1999, 291–301, 307–312, 316–326, 375–376, 390–391; Bell 2000; Hesseling 2003; and the essays in Shapiro 2005, 356–411. For a selection of Brouwer's writings, see Benacerraf and Putnam 1964, 66–84; Ewald 1996, 2:1166–1207; Mancosu 1998, 1–04, and Gray 2008, 200–304, 413–424, which admirably contextualizes mathematics in the larger currents of modernity.

[9] Other arguments have shown that $\sqrt{2}^{\sqrt{2}}$ is in fact irrational, but they are difficult and no easy proof is known. See Gel'fond 1960.

[10] Wittgenstein 1975, §§145, 173–174. For his reaction to Brouwer, see Hesseling 2003, 190–198, who argues that it was "the foundational battle between Brouwer and Hilbert which stimulated Wittgenstein to hold his later views." (197) For Wittgenstein's reactions to Weyl's work, see Hesseling 2003, 195.

[11] For Weyl's work on gauge theories, see Weyl 2009a, 3-5, Ryckman 2003, Fogel 2008, 45–120.

[12] See Weyl 1994 and Scholz 1994, 1995, 2000; for a contemporary reconsideration (and even vindication) of this work, see Feferman 1998, 51–58, 249–283; Feferman 2000, and his article in Shapiro 2005, 590–624.

[13] For a helpful selections of Hilbert's writings, see Benacerraf and Putnam 1964, 134–151; Ewald 1996, 2: 1087–1165; and Mancosu 1998, 149–274.

[14] See below, 17–19.

[15] See below, 21. Weyl refers, in a footnote below (83), to Ernst Cassirer, whose magisterial work on symbolic forms [1953-1996] are comparable to Weyl's preoccupations; for his critical commentary, see Weyl 2009b, 149–150, 216. see also Weyl 2009a, 194–195. For discussion of Weyl's symbolic constructivism, see Scholz 2006a, 2011b, and Toader 2011.

[16] See below, 28, 30. For Weyl's relation to Husserl, see Van Dalen 1984; Tonietti 1988; Feist 2002; Hesseling 2003, 106–107, 124–132; Tieszen 2005, 248–275; Ryckman 2007; Sieroka 2007; Gray 2008, 204–210; Mancosu 2011, 259–345; and Toader 2011. Meister Eckhart was an early fourteenth century mystical theologian who distinguished between God and the godhead, as a source of freely overflowing creation. Eckhart's teachings were accused of heresy by the medieval Church.

[17] For examples of Klein's pedagogical writings, see Klein 1939 and Pesic 2007, 109–116; regarding Klein's pedagogical work, see Tobies 1981. For Weyl's 1930 eulogy of Klein, see *WGA* 3:292–299.

[18] See below, 33, 47.

[19] For the fight between angel and devil, see Weyl 1939b, 681; for the Göttingen community, see Sigurdsson 1994, 1996, 2007; for Weyl's relations with Gödel, see Dawson 1997, 202.

[20] Einstein's comment was in a letter to the *New York Times* on 3 May 1935, included (with Weyl's address) in Catt 1935 and Dick 1981, 92–94; for Weyl's quote, see below 65.

[21] Reid 1996, 143.

[22] See below, 56. Roquette 2008 gives a helpful picture of the relations between Weyl and Noether. For Noether's influence, see McLarty 2006.

[23] Weyl also gave special courses of lectures at Yale (*The Open World*, 1932) and Swarthmore (*Mind and Nature*, 1934) and addressed the Princeton (1946) and Columbia (1954) bicentennial conferences; these are all included in Weyl 2009a.

[24] See below, 67.

[25] See below, 69, 70, 72.

[26] Reid 1996, 200–201.

[27] See below, 88.

[28] See below, 97, 113.

[29] See below, 117.

[30] Because Weyl's actual review [1946b] discusses in detail the various articles in a volume devoted to Russell's work [Schilpp 1951], it seemed preferable only to include here Weyl's self-sufficient preface [1946a].

[31] For this diagram, see below, 144.

[32] For this photo, see Weyl 2009a, 159.

[33] Weyl 2009a, 168 records his feelings about the role of relativity theory in his life; for the Weyl-Cartan arguments, see Pesic and Boughn 2003. See also the correspondence in Cartan and Einstein 1979; for Weyl's quote about himself, see below, 152–153.

[34] See below, 159, 170–172, 174, 188.

[35] See below 196–197, 202.

[36] This lecture is listed last on the chronological inventory in the ETH Archiv of the manuscripts in the Weyl *Nachlaß*. To the best of my knowledge, no other writing of his later than this lecture has come to light. I thank Christine Di Bella, Archivist at the Shelby White and Leon Levy Archives Center at the Institute for Advanced Study, Princeton, for her invaluable help in searching their records.

[37] Quoted from 211, below. Weyl first stated his speculations on Maxwell's equations limiting the dimensionality of space-time in the third (1919) edition of *Space-Time-Matter* (Weyl 1952a, 284); his treatment of Riemann surfaces is Weyl 2009c. For the history of conformal invariance, see Kastrup 2008. For the

earliest statements about the conformal invariance of Maxwell's equations, see Bateman 1909, 1910a, 1910b, and Cunningham 1910, though I have found no evidence that Weyl knew these papers.

[38] For the history of speculations about dimensionality, see Whitrow 1955 and Barrow 1983. Ehrenfest 1918 independently argued for the three-dimensionality of space based on the stability of dynamical orbits, which he connected with Weyl's work in his 1920 paper. Weyl does not mention Ehrenfest's work, though he may well have been aware of it via Einstein, a mutual friend.

[39] Many aspects of this field are in rapid development and change; see, for instance, Lämmerzahl and Macias 1993, Tegmark 1997, Nelson and Sakellariadou 2009.

[40] Bott 1988; Dyson 1956, 457. For views of Weyl's mathematical legacy, see Wells 1988 and Tent 2008.

Levels of Infinity
(1930)

Mathematics is the science of the infinite. The great achievement of the Greeks was to have made the tension between the finite and the infinite fruitful for the knowledge of reality. The feeling of the calm and unquestioning acknowledgement of the infinite belongs to the Orient, but for the East it remained a mere abstract awareness that left the concrete manifold of existence lying indifferently to one side, unshaped and impervious. Coming out of the Orient, the religious feeling of the infinite, the *apeiron*, took possession of the Greek soul in the Dionysian-Orphic epoch that preceded the Persian Wars. Here the Persian Wars also mark the release of the Occident from the Orient.[1] That tension and its overcoming became for the Greeks the driving motive of knowledge. But every synthesis, as soon as it was achieved, permitted the old antithesis to break out anew in deepened form. Thus, it determined the history of theoretical knowledge into our time. Indeed, today we are compelled everywhere in the foundations of mathematics to return directly to the Greeks.

Anaxagoras was the first to give the concept of the infinite a formulation that allowed it to play a role in science. A fragment that has come down to us from him says: "*In the small there is no smallest, but there is always a smaller. For what is can by no division, however far it is carried out, ever cease to be.*" This concerns space or body. The continuum, he says, cannot be put together out of discrete elements, which would be "chopped off from each other as if by an axe."[2] Space is not only infinite in the sense that nowhere within it does one come to an end, but at every place it is, so to speak, infinite in the inward direction. A point may be more and more exactly fixed only through a process of division extending to infinity, from level to level. This stands in contrast with the calm, completed existence of space for intuition. For the quality that fills it, space is the principle of separation, is that which makes possible the variability of the qualitative, but also is *separation* and, at the same time, *contact*, continual coherence, so that no piece may be "chopped off, as if by an axe," from another. Therefore no actual spatial thing can ever be adequately given because it unfolds its "inner horizon" in a process of ever newer and more precise experiences that unfolds to infinity.[3] Thereby it appears impossible through any such process to posit the actual thing as *existent*, closed and completed in itself. So the problem of the continuum drives one to epistemological *idealism*: Leibniz, among others, testified that the search for an exit from the "labyrinth of the continuum" first led him to the conception of space and time as orderings of the phenomena. "From the fact that body does not permit itself to be dissolved mathematically into primary 'fundamental moments' ['indivisibles'] it follows immediately," said Leibniz, "that it [body] is nothing substantial, but rather an ideal construction, which denotes only a *possibility of parts*, but certainly nothing real."

Against Anaxagoras arose the strict atomism of Democritus, one of whose arguments against the limitless divisibility of bodies runs something like this: "It is said that division is possible; very well, let it have happened. What remains? No bodies, for these could be further divided and their division would not have come to an end. Only points could remain, and body would have to consist of points, which is clearly absurd." The impossibility of grasping the continuum as a fixed being cannot be more powerfully illustrated than by the well-known paradox of Zeno about the race between Achilles and the tortoise. You know what it consists of: Assume that the tortoise has a head start of length 1 on Achilles; if Achilles is twice as fast as the tortoise, then at the moment when Achilles arrives where the tortoise started, the tortoise will be a distance $\frac{1}{2}$ ahead; after Achilles has covered this ground, the tortoise has gone a journey of length $\frac{1}{4}$; and so forth ad infinitum; hence it follows that swift-footed Achilles never catches up with the the reptile. The hint that the successive partial sums of the series

$$1 + \frac{1}{2} + \frac{1}{2^2} + \frac{1}{2^3} + \cdots$$

do not grow beyond all bounds but converge upon 2 —a hint by which today one may think to resolve the paradox — is certainly an important, relevant, and illuminating remark. But if the distance of length 2 actually consists of infinitely many partial distances of length $1, \frac{1}{2}, \frac{1}{4}, \frac{1}{8}, \ldots$ as "chopped off" wholes, then Achilles exhausting them all is contrary to the essence of infinity as "unfinishable." Hence Aristotle remarks on the solution of Zeno's paradox that "what is moved does not move by counting," or, more precisely, "when you divide the continuous line in two halves, you take the one [dividing] point for two; you make it both the beginning and the end. In dividing thus, neither the line nor the motion remains continuous.... In the continuous, there are indeed infinitely many halves, but they are not actually but *potentially*." Since Leibniz gives phenomena a foundation in a world of absolute substances, he too cannot evade Democritus's compelling argument and conceives the idea of the monad. "In the ideal or the continuum," said Leibniz in agreement with Aristotle, "the whole precedes the parts ... The parts are here only potential. In substantial things, however, the simple precedes the aggregate; the parts are actual and given before the whole. These reflections eliminate the difficulties regarding the continuum, difficulties that only arise when one sees the continuum as something real, which, prior to any division on our side, in itself possesses actual parts, and when one holds matter to be a substance."

The key word given here to solve the antinomy of the continuum is the distinction between actuality and potentiality, between *Being and Possibility*. Accordingly, in the end, the unfolding of the mathematical construction of reality rests on its subjective-objective double nature: it rests on the fact that it is not a being in itself, but rather an appearance for an intellectual I [*ein geistiges Ich*].[4] If, metaphysically, with Plato, one lets the picture that hovers before consciousness arise from the meeting of a "movement" from the I and one from the object, then one will make the I responsible for extension, the intuitional form of space and time, as the qualitatively undifferentiated field of free possibilities.[5] Mathematics is not the stiff and paralyzing schema the layman prefers to imagine; rather, with mathematics we stand precisely at that intersection of bondage and freedom that is the essence of the human itself.

If we now set ourselves to grasp these old ideas somewhat more precisely, we discover infinity first not in the continuum but instead in a still more primitive form in the *sequence of natural numbers* 1,2,3,...; only with their help can we approach at all the mathematical grasp of the continuum. In the development of arithmetic, as far as the role of infinity is concerned, four levels can be distinguished. To the *first* level belong such individual concrete judgments as $3 > 2$; the number-sign $||$ is a partial piece contained in the number-sign $|||$. To the *second* level belongs (for instance) the idea of $<$, of "being contained" for arbitrary number-signs and the judgment of *hypothetical universality*: given any two number-signs A, B, then either $A < B$, $A = B$, or, finally, $B < A$. The domain of the actually given is not exceeded here because this judgment only pronounces upon cases *in which* determinate numbers A, B are given. Something wholly new occurs, however, when at the *third* level I locate the actually occurring number-signs in the sequence of *all possible numbers* arising through a process of generation in accord with the principle that from a given number n, there can always be *generated* a new one, *the next* number, n'. *Here the existent is projected onto the background of the possible, of an ordered manifold of possibilities producible according to a fixed procedure and open to infinity.* Systematically, this standpoint is expressed in the definition and proof by complete induction: To establish that a property P belonging to some arbitrary natural number n belongs to every such number, it is enough to show that (a) 1 has the property P, (b) if n is any number with property P, then the next number n' also has property P. The distinction between even and odd numbers by the familiar "counting off by twos" is a simple example of definition through complete induction, which can be reduced to the form: (a) 1 is odd; (b) depending on whether n is even or odd, then n' is odd or even.[6]

The general propositions of number theory at this level concern the freedom of a self-developing sequence of numbers to stop at an arbitrary place. The transition to authentic theoretical knowledge is thereby accomplished: the transition from the *description a posteriori of the actually given* to the *construction a priori of the possible*, in whose manifold the given is assigned its ordered place not on the basis of descriptive features, but on the basis of the results of certain mental or physical manipulations and reactions to be carried out on it, for example the counting process. The foes of this method are on the one hand the empiricists (because a priori construction is a thorn in their flesh), who fondly imagine they can grasp Being as a single layer, unadorned, "by pure description" (Bacon contra Galileo, Hume contra Kant, Mach contra Einstein).[7] On the other hand there are the metaphysicians, who, out of hatred for freedom or for the field of construction insofar as it is open to infinity, build up a stiff dialectical concept-world as true Being (Hegel contra Newton). — We will only later address in more detail the fourth level of arithmetic, in which according to the model of the Platonic theory of ideas the *Possible* is turned into a transcendent and absolute *Being*, which is by nature inaccessible in its totality to open-eyed [*schauenden*] insight. For now, we will refrain from this dangerous step and turn from the natural numbers to the continuum, asking how we are to describe this substrate of possible divisions that proliferate ad infinitum.

As an example, I choose the one-dimensional closed continuum, say the perimeter of a circle. Let it first be divided into two pieces or arcs that border each other at their ends, A, B (the zeroth-level division). For the transition to the first-level division, each of these arcs is divided into two parts; for the second level, the same thing is repeated with each of the four sub-arcs that have thus arisen, etc. The number of arcs is multiplied by 2 at every step. (See the figure, in which the new divisions for every level are marked by crosses.)

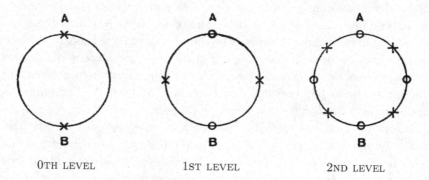

OTH LEVEL   1ST LEVEL   2ND LEVEL

In the succession of divisions for the zeroth, first, second, ..., nth, ... levels, the infinite series of numbers emerges. I would gladly stipulate that at every level the pieces that arose at the previous one are exactly halved, but I may not. For it is of the essence of the continuum that (1) it is divisible, yet (2) it is not divisible exactly (has no parts "chopped off from each other by an axe"), though (3) the fineness and precision of the division need nowhere find a limit at which to cease. Hence, successive divisions can only be carried out with a certain imprecision; but while the division proceeds from level to level, those vaguely determined limits established by the earlier divisions must become more and more precisely fixed. It is senseless to imagine that this process, which is virtually infinite, could have reached its end in an actual continuum; rather, it has always only reached a certain level. But we must keep the possibility open that the process can be taken a step further at every level. Hence, we separate the "arithmetically empty form [*Leerform*]" of the process of division from its actualization in a given continuum; the description of the empty form consists in specifying how the pieces (designated by appropriate symbols) of the nth level border one another and which of these pieces arise out of which pieces of the previous level by the nth division. The empty form, in contrast to its concrete realization, is defined a priori to infinity. The mathematical theory of the continuum is about this empty form. — We already know another arithmetically empty form: the sequence of numbers. Something undivided gets divided into one piece (the 1), now kept as a unit, leaving an undivided remainder; this remainder is divided again into a piece (2) and an undivided remainder, etc. (The most obvious manifestation of this: Of the time remaining open to us in the future, an ever new piece is always being lived through.) In this case, each part does not fall victim to division by two, only the last remaining undivided part. This schema is simpler than the schema of the division of the continuum, but in principle is of the same kind; in keeping with an ancient insight of Plato, the bifurcation of the One into the Dyad is the basis of both.[8]

Let us bring together every two of those contiguous arc-segments arising from the nth level division of the circumference as a *dual interval of the nth level*. These overlap one another so that if one knows a point with sufficient exactitude one may name with certainty the dual intervals of the nth level in which it lies. The particular place will then be more and more precisely determined, "trapped" by an infinite sequence of dual intervals of increasingly higher levels, each entirely enclosed within the previous level. This procedure is in principle equivalent to that whereby in antiquity Eudoxus taught how to trap locations in the continuum and to separate them from each other. Modernity has added the insight that one need not regard the sequence as a mere means for the conceptual description of a point given previously and assured to exist, but that through such a sequence the point in the continuum may be constructively generated in the first place. *Every* such sequence produces a point; in the arithmetically empty schema, the points are created precisely thereby. Only by means of this constructive turn is a mathematical mastery, an analysis, of continuity possible.

Yet there is still a problem with the idea of the infinite sequence. Everything essential is preserved, but the relationships are somewhat simpler to describe if instead of the schema of continued halving we begin with the schema of the natural numbers. A nascent sequence of natural numbers is established by my arbitrarily choosing the first number, then a second, third, etc.: *free choice sequence*. For *any arbitrary* sequence, for an *arbitrary* point, we deal with this free choice sequence. But it is only meaningful to attribute to the sequence those things that will be decided upon at some finite place in the development (e.g.,

"Does 1 occur among the numbers of the sequence within the first ten places?" rather than "Does 1 occur at all?" After all, the sequence is never finished!) A particular and determinate (indeed, determined to infinity) sequence cannot come out of a particular choice. Rather, a *law* is necessary that allows us to calculate from the arbitrary natural number n the number of the sequence that stands in the nth place. If one adheres to this schema, one sees that, in agreement with Aristotle's remark above, it is wholly impossible to take the continuous extension $0 \ldots 1$ apart into two pieces $0 \ldots \frac{1}{2}, \frac{1}{2} \ldots 1$ so that every place x either belongs to one half or the other.

Thus we have become acquainted with infinity in two forms: (1) the free possibility of stopping the sequence of numbers $1,2,3,\ldots$ at any arbitrary place (a particular act of choice); (2) the free possibility of constructing an emergent, unending sequence of natural numbers (an act of choice repeated ad infinitum) — which, however, if specified as a determinate sequence extending to infinity, turns into a law.

Before the discovery of the irrational by Pythagoras or the mathematicians of the Pythagorean school, as long as measure-numbers applied to extensions only resulted in whole-number ratios, the opinion prevailed that a particular point in the continuum could be determined by one (or two) natural numbers, that form (2) of infinity could be reduced to form (1). That would mean, however, that an arbitrary sequence of numbers Z through a lawful rule would determine a natural number n_z unambiguously, as if naming it. This is obviously impossible. For the name n_z must be determined when the sequence Z reaches a certain term. This need not be a fixed term nameable in advance, say the second or the hundred thousandth, but could depend on the outcome of the choice; but in the end, n_z must be settled and then can no longer be altered by further terms in the sequence. But then all sequences that are the same up to this term, as much as they may differ from one another past it, will deliver the same n_z.

So far I have largely followed the Dutch mathematician Brouwer, who in our day has strictly followed out the intuitionist standpoint in mathematics, which is characterized by the tension between being and possibility.

Metaphysics has always attempted to overcome the dualism of object and subject, being and possibility, being and meaning, bondage and freedom. In *realism*, the object, and in *idealism*, the subject was one-sidedly exalted to the status of absolute Being, as most powerfully expressed by the early Fichte.[9] In *theism* — the contrary direction of transcendence from realism — God is posited as absolute Being, flowing out of whom the light of consciousness, whose origin itself is inaccessible, grasps itself in its full penetration of its own being as existence in the tension and split between meaning and being. Mathematics, too, from its own motive or from its dependence on philosophy, has not withstood the pull toward the absolute. Now that in the first part I have shown you the mathematical infinite as possibility, I will speak in the second part of the attempts to turn the field of possibilities opening to infinity into a closed realm of absolute existence. Four different attempts to reach this goal have emerged in the course of history, though only the first two refer to the continuum.

The first and most radical has the continuum consist of countable, discrete, indivisible elements: atoms. For matter, this path, already trodden in antiquity by Democritus, has been traversed with the most dazzling success by modern physics. For space itself, Plato seems to have been the first to have laid out a consistent atomism — with clear awareness of the goal he had set: the "saving" of the appearances by the ideas. The atomistic theory of space was renewed in Islamic philosophy by the Mutakallimūn, in Europe by Giordano Bruno's doctrine of the minimum.[10] Stimulated by quantum theory, this thought turns up in our time once again in discussions on the foundations of physics. But up to now it has always been mere speculation, has remained stuck at its beginning, and has never achieved the least contact with reality.

The second attempt is the infinitely small. The tangent of a curve at point P_0 is treated as a straight line connecting P_0 to an infinitely nearby point on the curve, not as the ultimate location that the secant $\overline{P_0 P}$ approaches without limit as the point P on the curve converges to P_0; the velocity is the quotient of the infinitely small distance covered in the infinitely small time dt divided by dt, not the limit to which the analogously constructed quotient for a finite time interval converges when one shrinks the time interval smaller than any bounds. Galileo compares the breaking up of a straight line into a regular chiliagon to its rolling up into a circle: breaking up into an infinitely many-sided polygon of infinitely small sides is here, according to Galileo, actually accomplished, though it is not possible to separate the sides individually one from another. He understands condensation and rarefaction as a mixture of infinitesimally filled and empty parts of space in changing proportions. Johann Bernoulli employs the seemingly unanswerable argument regarding the sequence $\frac{1}{2} + \frac{1}{4} + \frac{1}{8} + \cdots$ that if there are ten members, take the tenth; if there are one hundred, take the hundredth; if infinitely many, then take that one. Hence, says Bernoulli, this sequence possesses an infinitely manyth member, which is naturally infinitesimal. Leibniz, meanwhile holds fast to the thought that every determinate member of the sequence is a definite, *finite* fraction. Bernoulli would be right if the sequence with all of its members could be represented as complete. While Eudoxus had cast the infinitesimal aside with a pointed axiom, in the eighteenth century this very idea, vague and incomprehensible, becomes the foundation of infinitesimal calculus. "The incomprehensibilities of mathematics" is a favorite expression from the beginning of the century. The founders, indeed, Newton and Leibniz, had enunciated relatively clearly the right understanding, that it was not a matter of a fixed infinitesimal but rather of crossing the boundary to zero; but this standpoint does not dominate the superstructure of their thought, and they obviously do not know that carrying out the limit process must not only determine the [numerical] value of the limit but must first guarantee its existence. Thus, for centuries the modern infinitesimal calculus could not withstand comparison in logical rigor with the Greek theory of the continuum. At the same time, the reach of the problem had grown far greater; it is from the start a question of the analysis of arbitrary, continuous shapes and processes, especially processes of motion. The passionate will to reality is stronger in our culture than the clear-sighted Greek *ratio*.[11]

The limit process at last carried the day and thereby shattered also the second attempt to confine the continuum of Becoming into a frozen Being. For limit is an unavoidable concept whose importance is not affected by the assumption or dismissal of the infinitely small. Once it has been grasped, one sees that it renders the infinitely small superfluous. Starting with the behavior of the infinitely small as governed by elementary laws, infinitesimal analysis wants to draw conclusions, by means of integration, about the behavior of the finite, for instance, from the universal law of attraction for two mass-filled "volume elements" to the magnitude of the attractive force of arbitrarily shaped, homogeneous or non-homogeneous extended bodies endowed with mass. But unless one interprets the infinitesimal here as "potential" in the sense of the limit process, then the one has nothing to do with the other, the processes in the finite and the infinitesimal become entirely independent of each other and the connecting bond is cut. Here, Eudoxus undoubtedly had the right view.

At first glance, it might appear as if this victory of the limit process were the final realization of Aristotle's teaching that the infinite exists only *dynamei*, according to potentiality, in nascence and evanescence but not *energeiai* [according to actuality]. An error! In the nineteenth century from Cauchy to Weierstrass, grounding analysis especially on the limit concept issued in a new, powerful attempt to overcome the dynamics of infinity to the benefit of the static: *set theory*. Any particular convergent series, such as, for instance, the series of partial sums of the Leibnizian series

$$\frac{1}{1} - \frac{1}{3} + \frac{1}{5} - \frac{1}{7} + \cdots,$$

which converges to $\frac{\pi}{4}$, does not develop in a lawless process to which we must blindly entrust ourselves to see what it brings forth at each step but is fixed once and for all by a determinate *law* assigning every natural number n its own value within the approximation, the nth partial sum. But law is static. And if we ask what it means that the sequence of points $P_1, P_2, \ldots, P_n, \ldots$ converges toward point P, analysis answers us with the explanation: it means that there exists for *every* positive interval ϵ a natural number $N = N(\epsilon)$ such that the distance $\overline{PP_n}$ is smaller than ϵ for *all* $n \geq N(\epsilon)$. The dynamics of the transition to the limit are reflected here again in a static relation between the sequence $\{P_n\}$ and the point P, a relation that can admittedly only be grasped by the *unlimited use of the expressions "there exists" and "all" about the series of natural numbers*. This is the standpoint we had earlier posited as the fourth level of arithmetic. Whoever accepts the definition that appeals to the infinite totality of numbers, "n is an even or odd number, according to whether there exists a number x for which it is true that $n = 2x$ or not," for him, the sequence of numbers opening to infinity has transformed itself into a closed [*Inbegriff*] realm of things existing in themselves, a realm of absolute existence that is "not of this world" and from which only bit by bit a reflection arrives for the gaze of our consciousness. In that absolute realm, for any property P that is predicable of numbers, the *tertium non datur* ["a third possibility is not given"] holds, i.e., the alternative: either there exists a number with property P, or all numbers have the contrary property not-P. This could only be determined in all cases, however, if one could scrutinize the whole sequence of numbers with a view to property P, but that would contradict the essence of the infinite. We are therefore prevented from interpreting the statement of existence as an infinite logical sum[12] "1 has the property P, or 2 or 3, ad infinitum." We are prevented from understanding the universal statement as the infinite logical product "1 has the property not-P, and 2 and 3, ad infinitum." One must understand the universal statement that treats "every number" *hypothetically,* i.e. as saying something only assuming a number is really given and so this statement is not capable of negation. And the statement of existence only acquires a content in regard to the example offered: this determinate number, constructed in such and such a way, has the property P. Existential absolutism, however, sets itself above and beyond these concerns based on the essence of the infinite and takes those statements as ordinary and capable of negation and as opposable to one another in the *tertium non datur* judgment.

Indeed, set theory proceeds even more radically: it uses the expressions "*there exists*" and "*for all*" unrestrainedly with possible sequences or sets of natural numbers — under the thinking that such sequences refer to a state of affairs that is determined in the things themselves by Yes or No, whether or not our insight can succeed by happy chance of mathematical method in transforming this mute and closed answer into an explicit one open to the light of day. We speak of the set of all even numbers, of the set of all prime numbers. A set is hence always described as all numbers with such and such a property; the set counts as given if a determinate criterion decides which elements belong to it and which do not. Similarly, an infinite sequence can only be given by a determinate law that fixes the number in the nth place by its dependence on the place designator n. But if it be asked whether among all possible sets or sequences there exists a number of such and such kind or not, one can scarcely rid oneself of the feeling that through the attachment of "accessible" sets and sequences to the law, a chaotic plenum of possibilities, of "arbitrarily tossed together," "lawless" sets is neglected and that thereby the clear alternative, "Does it exist or not?" get confused. Set theory is untroubled by its use of such alternatives in the criteria it erects to determine the membership of a number in a set or in the laws by which it defines an infinite sequence. One can see that it thereby entangles itself in a pernicious logical circle. Admittedly no genuine contradictions in analysis have resulted thus far; we do not completely understand why that is. G. Cantor, however, threw off all fetters, using the concept of the set with complete freedom, especially when he allowed the construction of the set of all subsets of a given set. Here, one first struck real contradictions on the outermost borders of set theory. Their root can only be uncovered as the daring act already committed from the beginning of mathematics on: a field of constructive possibilities was treated as a closed realm of things existing in themselves.

The method of set theory has taken over not only analysis but also arithmetic, indeed the origin of mathematics, the theory of natural numbers, and has promised to reduce them to general logical concepts like "there exists," "all," "one-to-one correspondence." Perhaps the method can best be explained by an example from this domain. The sequence of natural numbers appears to the set theorist as a complete sequence Z within which a map from $n \to n'$ is defined that assigns to every element n of the set a unique determinate element n' (the number following n in the sequence). Dedekind names a set of numbers C a *chain* if for every arbitrary element n of C, there also occurs n' in C. The property of a number n to be ≥ 5 can then be formulated: that number is contained in every chain in which 5 occurs. The finite criterion ("if the counting of the numbers from 1 to n goes through 5") is here replaced by a transfinite criterion, which, according to its wording, requires the inspection of *all possible subsets of Z*. But instead of something specifically arithmetical, "always one more," repetition ad infinitum, general logical concepts step forth. — For set theory, in principle there is no fundamental boundary between the finite and the infinite; infinity even seems the simpler. The diversity of finite sets, according to their size, there corresponds a much richer scale of different levels of infinity, as G. Cantor has shown in his fundamental studies of abstract set theory.

But we can hardly still believe today that something comprehensible is to be found behind these theories of Cantor. At least the differences between infinite cardinal numbers will be relative to the means of construction, with the help of which sets, sets of sets, . . . may be constructed in a given domain, so that in any case they forfeit their universal and absolute meaning for things altogether. Thanks to the critique of H. Poincaré, B. Russell, Brouwer, Skolem, and others, we have gradually had our eyes opened to the untenable logical positions from which the method of set theory proceeds. In my opinion, there can be no further doubt that this third attempt too — in the sense which it was intended — has foundered. A rescue of mathematics without diminution of its classical integrity is only possible, as D. Hilbert first recognized, through a radically changed interpretation of its meaning, namely through a formalization in which, in principle, it changes from a system of knowledge gained through insight into a game carried out according to fixed rules, with signs and formulas. This is the fourth and last attempt to overcome the tension between being and possibility that we have to address. As the usual representation by symbols in mathematics is extended also to the logical operations "and," "or," "there exists," etc., every mathematical statement is transformed into a meaningless formula built out of signs, and mathematics itself transforms into a game with formulas regulated by certain conventions, quite comparable to chess. To the chessmen correspond a limited or unlimited store of signs in mathematics, to an arbitrary layout of pieces on the board corresponds the combination of the signs in a formula. One or several formulas count as axioms; their counterpart is the prescribed arrangement of chessmen at the beginning of a game. And just as the later arrangement of pieces in the game arises out of the previous one through a move that must satisfy the rules for moving, likewise formal rules of inference reign, according to which new formulas can be produced, i.e. "deduced" out of previous formulas. In chess, I understand a permissible arrangement of pieces as one that arises from the starting arrangement in the course of a game played according to the rules of movement. The analogue in mathematics is the provable (or better, the proven) formula, which arises from the axioms on the basis of the rules of syllogism. Certain formulas of clearly described character are branded as contradictions; in chess we understand as a contradiction, say, every arrangement of pieces in which more than eight pawns of the same color appear. Formulas of other kinds stimulate those who play at mathematics, just as a chess problem proposing checkmate in so many moves stimulates the chess player. The mathematician seeks a skillful linking of moves into a final formula that will let him win the proof-game by proper play. Up to this point, everything is a game, not knowledge; however, now the game is made into the object of knowledge in "*metamathematics*," as Hilbert puts it: it should be acknowledged that a contradiction can never appear as the final formula of a proof. Hilbert wants to establish with certainty only this *freedom from contradiction*, not the truth of the content of the analysis. Analogously, it is no longer a game but knowledge when one recognizes that in chess more than eight pawns of the same color cannot occur. One recognizes it thus: At the beginning there are eight pawns; the number of pawns can never be increased by a move according to the rules for moving; ergo ... This "ergo" stands for a proof by complete induction, which follows the moves of the given game, step by step, to the final arrangement. Hilbert requires thinking with content and meaning only to acquire this one piece of knowledge; his proof of non-contradiction proceeds in principle like the one just carried out for the chess game, though naturally it is much more complicated. It goes without saying that these reflections on content confine themselves entirely within the limits of content-thinking laid down by Brouwer.

From this formalistic standpoint, one is no more permitted to seek deeper grounds for the axioms and rules of operation that have been assumed than one would be for a chess game; indeed, it even remains obscure why we are concerned that the game be "non-contradictory." The floor is closed to all objections, since indeed nothing is being asserted; refusal could only be expressed in the declaration "I am not joining in the game." If for the sake of preserving its certainty mathematics wished in all seriousness to withdraw to this position of mere play, it would thereby pass entirely out of the world-history of the mind. Therefore we must after all attempt somehow to assign a role to mathematics in the service of knowledge. Hilbert himself says somewhat obscurely that infinity plays the role of an idea in the Kantian sense, by which the concrete is completed in the sense of totality. I understand that to mean something like the way in which I complete what is given to me as the actual content of my consciousness, into the totality of an objective world, which certainly includes much that is not present to me. The scientific formulation of this objective concept of the world occurs in physics, which avails itself of mathematics as a means of construction. However, the situation we find before us in theoretical physics in no way corresponds to Brouwer's ideal of a science. That ideal postulates that every judgment has its own meaning achievable in intuition. The statements and laws of physics, nevertheless, taken one by one, have no content verifiable in experience; only the theoretical system as a whole allows itself to be confronted by experience. What is accomplished here is not the intuitive insight into singular or general contents and a description that truly renders what is given, but instead a theoretical, and ultimately purely symbolic, construction of the world. It has been said that physics deals only with the ascertaining of coincidences; Mach in particular has spoken for a pure phenomenalism in the field of physics. But if we are honest, we must after all admit that our theoretical interest does not depend exclusively or even primarily on "real statements," on the confirmation that this indicator is at this point on a scale, but rather on ideal positings that, according to theory, identify themselves in such coincidences, but whose meaning is not immediately fulfilled in any intuition that may emerge — such as, for example, the positing of the electron as a universal electrical elementary quantum. It cannot be denied that there lives in us a need for theory that is absolutely inexplicable from a merely phenomenalist standpoint. That need, which has a creative urge directed to a symbolic shaping of the transcendent, demands satisfaction and is driven by the metaphysical belief in the reality of the outer world (next to which the belief in the reality of one's own I, of the stranger's "thou," and of God takes its place in the same way).

It is a deep philosophical question what is the "truth" or objectivity corresponding to this theoretical shaping of the world that presses so far beyond the given. One indispensable requirement is, in any case, the *agreement* that can somewhat coarsely be expressed as follows: all possible indirect determinations of a physical magnitude on the basis of the theory must lead to the same result. This demand implies the non-contradictory character of the theory, so that we also receive here a reasonable answer to the question why we are concerned with the formal character of non-contradiction in our theory: because that aspect of the agreement pertains to the theory alone, not yet confronted with reality. The task of the mathematician is to see to it that the theories in the concrete sciences satisfy this *conditio sine qua non* that they be formally determinate and non-contradictory.

The development of science has clearly shown that various theoretical constructions of the world can satisfy the postulate of agreement. What brings about a decision between two such competing theories, a decision that will seem compelling to virtually all those who engage in that science, is not easy to say. Here we are obviously carried by the life-process of the mind accomplishing itself in us and we are not in a position to extract its achievement in the form of lifeless, final "results." The epistemological reflections merely adumbrated here lead me in any case to the following position. If one takes mathematics by itself, one should restrict oneself with Brouwer to the truths of insight, in which infinity only enters as an open field of possibilities; there is no motive discoverable that presses farther than that. But in natural science, we touch a sphere impervious anyhow to the demands of open-eyed [*schauenden*] certainty. Here, knowledge necessarily becomes symbolic shaping, which is why when mathematics is taken along by physics in the process of theoretical world-construction it is no longer necessary that the mathematical let itself be isolated as a particular region of the intuitively certain. From this lofty standpoint, whence all science appears as a unity, I grant Hilbert's point.

I will attempt in conclusion to gather up in a few general theses the experience that mathematics has acquired in the course of its history by means of its researches into infinity.

1) In the life of the human spirit, there are two clearly separate realms: one of *action*, of shaping, of construction, to which the active artist, scientist, technician, statesman is devoted and which in the area of science stands under the norm of objectivity — the other is a realm of *reflection*, which fulfills itself in insights and which one may regard by contrast as the proper domain of the philosopher. The danger of creative activity not watched over by reflection is that it escapes meaning, wanders, sinks to sclerotic routine — the danger of reflection, that it becomes a stumbling block to the creative force of the human, a "talking about things" that lacks any authority. What we were pursuing here was reflection. Hilbert's mathematics, like physics, belongs in the domain of constructive action; metamathematics, with its insight of non-contradiction, belongs to reflection.

2) The task of knowledge can certainly not be achieved by open-eyed [*schauende*] insight alone in that place where, as in natural science, we make contact with an objective sphere whose origin is impenetrable to reason. But even in pure mathematics, indeed in pure logic, we cannot perceive the validity of a formula by means of seeing in it a descriptive identifying feature; instead, such validity produces itself only by practical action, when we begin from the axioms and apply repeatedly and in arbitrary combination the practical rules of inference. In this sense, one can speak of an original darkness of reason: we do not possess the truth; it is not enough to open our eyes wider, for the truth is to be won by action.[13]

3) The infinite is accessible to mind and intuition in the form of a field of possibilities opening to infinity, as with the always further continuable sequence of numbers; but

4) The completed, actual infinite as a closed realm of absolute existence cannot be given to the mind.

5) Nevertheless, mind is ineluctably compelled by the demand of totality and the metaphysical belief in reality to represent infinity as closed Being by means of a symbolic construction.

I take these experiences drawn from the development of mathematics with philosophical seriousness. If you will allow me a theological version, then the last three points would say: We deny the thesis of the absolute finitude of the human being, both in the atheistic form of obdurate finitude and in the theistic form where it serves as the basis for the violent drama of crushing despair, revelation, and mercy. Instead, spirit is freedom in the bondage of existence, it is open to infinity. God as the completed infinite can admittedly not be given to us nor come to be given; God can neither break into the human through revelation nor can the human by mystic contemplation break through to Him. We can only represent the completed infinite in the symbol. All creative shaping by humans receives its deep holiness and dignity from this connection. Yet up to now it has been only in mathematics and physics, as far as I can see, that the symbolic-theoretical construction has taken on such solidity that it is compelling for everyone whose mind opens itself to these sciences.

Notes

[Translated from Weyl 1931 (not included in *WGA*), by Cary Stickney, edited and revised by Peter Pesic, with special thanks to Philip Bartok, Brandon Fogel, and Norman Sieroka for their invaluable criticism and suggestions. The title page of the original edition (1931) notes that this lecture was given in Jena on "27 October 1930 at the opening of the visitor's conference of the Mathematical Society of the University of Jena in the Abbeanum," a building that had just been built in Bauhaus style for educational and research use.]

[1] [During 499-449 B.C.E., the Greeks repulsed two massive Persian invasions, which Weyl takes as emblematic of the parting of the ways between the East and the West. Note that the Greek word *a-peiron* literally means "un-bounded" or "un-limited," emphasizing the sense that the "in-finite" is potentially unbounded rather than an actually completed being.]

[2] [Literally, *apokekoptai pelekei*, "hacked apart with an axe," meaning a large double-edged weapon used for felling trees or opponents in battle; see Anaxagoras 2007, 20-21, 52.]

[3] [Weyl explores and explains the unfolding of the "inner horizon" in 2009a, 29-33; he discusses the theme of openness in *The Open World*, included in Weyl 2009a, 34-82.]

[4] [This literal English translation tries avoid the misleading connotations of the term "ego," with its psychological (especially Freudian) associations; note

that *geistiges* is not restricted to the religious connotations of "spiritual" but here may mean more nearly "conscious."]

[5] [See Plato, *Theaetetus* 154ff.]

[6] [It should be noted that Weyl's use of the term "complete induction" differs from current usage, in which this term requires in step (b) that *n* and all numbers less than *n* have the property *P*, whereas present-day "ordinary induction" only requires that *n* have the property *P*, as Weyl assumes in what he calls "complete induction."]

[7] [Weyl's phrase *ohne Zutat* means literally "without trimming, ornamentation, seasoning, garnish," here translated as "unadorned."]

[8] [For discussion of Plato's treatment of the One and the Dyad, see Klein 1985, 285–288, and Reale 1990, 2:14, 65–73.]

[9] [For Weyl's relation to Fichte, see Weyl 2009a, 214–219, and Sieroka 2007, 2009, 2010, 2012.]

[10] [In the Muslim world during the eighth through twelfth centuries, the Mu'tazilī *mutakallimūn*, the philosophers of *kalām* (meaning speculative theology), argued that not only matter but also space was atomic; see Wolfson 1976 and Pesic 2002, 26–28.]

[11] [The meanings of the Greek word *logos* include "word," "reason," and "mathematical ratio"; this spectrum of meanings also informs its translation into the Latin word *ratio*.]

[12] I analogize the operation "or" to the arithmetic $+$, "and" to $\times$, as is usual in mathematical logic since Leibniz.

[13] [Weyl here may be alluding to Socrates' habit of opening his eyes very wide to express wonder or to underline the need to see something clearly. See, for instance, Plato, *Phaedo* 86d: "Socrates opened his eyes very wide [*diablepsas*] — a favorite trick of his — and smiled." As the Greek verb *diablepein* means "to stare with eyes wide open, to see clearly," the German verb *schauen* has something of this sense of open-eyed gazing, which our translation tries to capture throughout. Weyl may here also refer critically to the Husserlian concept of *Wesensschau*, "gazing at essences," meaning the grasping of the essential aspects of the phenomenological structure of cognition through a kind of inward "looking with eyes wide open," which Weyl judged to be insufficiently active.]

Topology and Abstract Algebra as Two Roads of Mathematical Comprehension

(1932)

We are not very pleased when we are forced to accept a mathematical truth by virtue of a complicated chain of formal conclusions and computations, which we traverse blindly, link by link, feeling our way by touch. We want first an overview of the aim and of the road; we want to understand the *idea* of the proof, the deeper context. A modern mathematical proof is not very different from a modern machine, or a modern test setup: the simple fundamental principles are hidden and almost invisible under a mass of technical details. When discussing Riemann in his lectures on the history of mathematics in the nineteenth century, Felix Klein said:

> Undoubtedly, the capstone of every mathematical theory is a convincing proof of all of its assertions. Undoubtedly, mathematics inculpates itself when it foregoes convincing proofs. But the mystery of brilliant productivity will always be the posing of new questions, the anticipation of new theorems that make accessible valuable results and connections. Without the creation of new viewpoints, without the statement of new aims, mathematics would soon exhaust itself in the rigor of its logical proofs and begin to stagnate as its substance vanishes. Thus, in a sense, mathematics has been most advanced by those who distinguished themselves by intuition rather than by rigorous proofs.

The key element of Klein's own method was an intuitive perception of inner connections and relations whose foundations are scattered. To some extent, he failed when it came to a concentrated and pointed logical effort. In his commemorative address for Dirichlet, Minkowski contrasted the minimum principle that Germans tend to name for Dirichlet (and that was actually applied most comprehensively by William Thomson) with the true Dirichlet principle: to conquer problems with a minimum of blind computation and a maximum of insightful thoughts. It was Dirichlet, said Minkowski, who ushered in the new era in the history of mathematics.[1]

What is the secret of such an understanding of mathematical matters, what does it consist in? Recently, there have been attempts in the philosophy of science to contrast understanding, the art of interpretation as the basis of the humanities, with scientific explanation, and the words intuition and understanding have been invested in this philosophy with a certain mystical halo, an intrinsic depth and immediacy. In mathematics, we prefer to look at things somewhat more soberly. I cannot enter into these matters here, and it strikes me as very difficult to give a precise analysis of the relevant mental acts. But at least I can single out, from the many characteristics of the process of understanding, one that is of decisive importance. One separates in a natural way the different aspects of a subject of mathematical investigation, makes each accessible through its own relatively narrow and easily surveyable group of assumptions, and returns to the complex whole by combining the appropriately specialized partial results. This last synthetic step is purely mechanical. The great art is in the first, analytic, step of appropriate separation and generalization. The mathematics of the last few decades has reveled in generalizations and formalizations. But to think that mathematics pursues generality for the sake of generality is to misunderstand the sound truth that a natural generalization *simplifies* by reducing the number of assumptions and by thus letting us understand certain aspects of a disarranged whole. Of course, it can happen that different directions of generalization enable us to understand different aspects of a particular concrete issue. Then it is subjective and dogmatic arbitrariness to speak of the true ground, the true source of an issue. Perhaps the only criterion of the naturalness of a severance and an associated generalization is their fruitfulness. If this process is systematized according to subject matter by a researcher with a measure of skill and "sensitive fingertips" who relies on all the analogies derived from his experience, then we arrive at axiomatics, which today is an instrument of concrete mathematical investigation rather than a method for the clarification and "deep-laying" of foundations.

In recent years mathematicians have had to focus on the general and on formalization to such an extent that, predictably, there have turned up many instances of cheap and easy generalizing for its own sake. Pólya has called it generalizing by dilution. It does not increase the essential mathematical substance. It is much like stretching a meal by thinning the soup. It is deterioration rather than improvement. The aged Klein said: "Mathematics looks to me like a store that sells weapons in peacetime. Its windows are replete with luxury items whose ingenious, artful and eye-catching execution delights the connoisseur. The true origin and purpose of these objects — the strike that defeats the enemy — have receded into the background and have been all but forgotten." There is perhaps more than a grain of truth in this indictment, but, on the whole, our generation regards this evaluation of its efforts as unjust.

There are two modes of understanding that have proved, in our time, to be especially penetrating and fruitful. The two are topology and abstract algebra.

A large part of mathematics bears the imprint of these two modes of thought. What this is attributable to can be made plausible at the outset by considering the central concept of real number. The system of real numbers is like a Janus head with two oppositely directed faces. In one respect it is the domain of the operations $+$ and $\times$ and their inverses, in another it is a continuous manifold, and the two are continuously related. One is the algebraic and the other is the topological face of numbers. Since modern axiomatics is simpleminded and (unlike modern politics) dislikes such ambiguous mixtures of peace and war, it made a clean break between the two. The notion of size of number, expressed in the relations $<$ and $>$, occupies a kind of intermediate relation between algebra and topology.

Investigations of continua are purely topological if they are restricted to just those properties and differences that are unchanged by arbitrary continuous deformations, by arbitrary continuous mappings. The mappings in question need only be faithful to the extent to which they do not collapse what is distinct. Thus it is a topological property of a surface to be closed like the surface of a sphere or open like the ordinary plane. A piece of the plane is said to be simply connected if, like the interior of a circle, it is partitioned by every crosscut. On the other hand, an annulus is doubly connected because there exists a crosscut that does not partition it but every subsequent crosscut does. Every closed curve on the surface of a sphere can be shrunk to a point by means of a continuous deformation, but this is not the case for a torus. Two closed curves in space can be intertwined or not. These are examples of topological properties or dispositions. They involve the primitive differences that underlie all finer differentiations of geometric figures. They are based on the single idea of continuous connection. References to a particular structure of a continuous manifold, such as a metric, are foreign to them. Other relevant concepts are limit, convergence of a sequence of points to a point, neighborhood, and continuous line.

After this preliminary sketch of topology I want to tell you briefly about the motives that have led to the development of abstract algebra. Then I will use a simple example to show how the same issue can be looked at from a topological and from an abstract-algebraic viewpoint.

All a pure algebraist can do with numbers is apply to them the four operations of addition, subtraction, multiplication, and division. If a system of numbers is a field, that is, if it is closed under these operations, then the algebraist has no means of going beyond it. The simplest field is the field of rationals. Another example is the field of numbers of the form $a + b\sqrt{2}, a, b$ rational. The well-known concept of irreducibility of polynomials is relative and depends on the field of coefficients of the polynomials, namely a polynomial $f(x)$ with coefficients in a field K is said to be irreducible over K if it cannot be written as a product $f_1(x) \cdot f_2(x)$ of two non-constant polynomials with coefficients in K. The solution of linear equations and the determination of the greatest common divisor of two polynomials by means of the Euclidean algorithm are carried out within the field of the coefficients of the equations and of the polynomials respectively. The classical problem of algebra is the solution of an algebraic equation $f(x) = 0$ with coefficients in a field K, say the field of rationals. If we know a root θ of the equation, then we know the numbers obtained by applying to θ and to the (presumably known) numbers in K the four algebraic operations. The resulting numbers form a field $K(\theta)$ that contains K. In $K(\theta), \theta$ plays a role of a determining number from which all other numbers in $K(\theta)$ are rationally derivable. But many — virtually all — numbers in $K(\theta)$ can play the same role as θ. It is therefore a breakthrough if we replace the study of the equation $f(x) = 0$ by the study of the field $K(\theta)$. By doing this we eliminate all manner of trivia and consider at the same time all equations that can be obtained from $f(x) = 0$ by means of Tschirnhausen transformations.[2] The algebraic, and above all the arithmetical, theory of number fields is one of the sublime creations of mathematics. From the viewpoint of the richness and depth of its results it is the most perfect such creation.

There are fields in algebra whose elements are not numbers. The polynomials in one variable, or indeterminate, x, [with coefficients in a field], are closed under addition, subtraction, and multiplication but not under division. Such a system of magnitudes is called an integral domain. The idea that the argument x is a variable that traverses continuously its values is foreign to algebra; it is just an indeterminate, an empty symbol that binds the coefficients of the polynomial into a uniform expression that makes it easier to remember the rules for addition and multiplication. 0 is the polynomial all of whose coefficients are 0 (not the polynomial which takes on the value 0 for all values of the variable x). It can be shown that the product of two nonzero polynomials is $\neq 0$. The algebraic viewpoint does not rule out the substitution for x of a number a taken from the field in which we operate. But we can also substitute for x a polynomial in one or more indeterminates $y, z, \ldots$. Such substitution is a formal process which effects a faithful projection of the integral domain $K[x]$ of polynomials in x onto K or onto the integral domain of polynomials $K[y, z, \ldots]$; here "faithful" means subject to the preservation of the relations established by addition and multiplication. It is this formal operating with polynomials that we are required to teach students studying algebra in school. If we form quotients of polynomials, then we obtain a field of rational functions which must be treated in the same formal manner. This, then, is a field whose elements are functions rather than numbers. Similarly, the polynomials and rational functions in two or three variables, x, y or x, y, z with coefficients in K form an integral domain and field respectively.

Compare the following three integral domains: the integers, the polynomials in x with rational coefficients, and the polynomials in x and y with rational coefficients. The Euclidean algorithm holds in the first two of these domains, and so we have the theorem: If a, b are two relatively prime elements, then there are elements p, q in the appropriate domain such that

(*) $$1 = p \cdot a + q \cdot b.$$

This implies that the two domains in question are unique factorization domains. The theorem (*) fails for polynomials in two variables. For example, $x - y$ and $x + y$ are relatively prime polynomials such that for every choice of polynomials $p(x, y)$ and $q(x, y)$ the constant term of the polynomial $p(x, y)(x-y)+q(x, y)(x+y)$ is 0 rather than 1. Nevertheless polynomials in two variables with coefficients in a field form a unique factorization domain. This example points to interesting similarities and differences.

There is yet another way of making fields in algebra. It involves neither numbers nor functions but congruences. Let p be a prime integer. Identify two integers if their difference is divisible by p, or, briefly, if they are congruent mod p. (To "see" what this means, wrap the real line around a circle of circumference p.) The result is a field with p elements. This representation is extremely useful in all of number theory. Consider, for example, the following theorem of Gauss that has numerous applications: If $f(x)$ and $g(x)$ are two polynomials with integer coefficients such that all coefficients of the product $f(x) \cdot g(x)$ are divisible by a prime p, then all coefficients of $f(x)$ or all coefficients of $g(x)$ are divisible by p. This is just the trivial theorem that the product of two polynomials can be 0 only if one of its factors is 0, applied to the field just described as the field of coefficients. This integral domain contains polynomials that are not 0 but vanish for all values of the argument; one such polynomial is $x^p - x$. In fact, by Fermat's theorem, we have

$$a^p - a \equiv 0 \pmod{p}.$$

Cauchy uses a similar approach to construct the complex numbers. He regards the imaginary unit i as an indeterminate and studies polynomials in i over the reals modulo $i^2 + 1$, that is he regards two polynomials as equal if their difference is divisible by $i^2 + 1$. In this way, the actually unsolvable equation $i^2 + 1 = 0$ is rendered, in some measure, solvable. Note that the polynomial $i^2 + 1$ is prime over the reals. Kronecker generalized Cauchy's construction as follows. Let K be a field and $p(x)$ a polynomial prime over K. Viewed modulo $p(x)$, the polynomials $f(x)$ with coefficients in K form a field (and not just an integral domain). From an algebraic viewpoint, this process is fully equivalent to the one described previously and can be thought of as the process of extending K to $K(\theta)$ by adjoining to K a root of the equation $p(\theta) = 0$. But it has the advantage that it takes place within pure algebra and gets around the demand for solving an equation that is actually unsolvable over K.

It is quite natural that these developments should have prompted a purely axiomatic buildup of algebra. A field is a system of objects, called numbers, closed under two operations, called addition and multiplication, that satisfy the usual axioms: both operations are associative and commutative, multiplication is distributive over addition, and both operations are uniquely invertible yielding subtraction and division respectively. If the axiom of invertibility of multiplication is left out, then the resulting system is called a ring. Now "field" no longer denotes, as before, a kind of sector of the continuum of real or complex numbers but a self-contained universe. One can apply the field operations to elements of the same field but not to elements of different fields. In this process we need not resort to artificial abstracting from the size relations $<$ and $>$. These relations are irrelevant for algebra and the "numbers" of an abstract "number field" are not subject to such relations. In place of the uniform number continuum of analysis we now have the infinite multiplicity of structurally different fields. The previously described processes, namely adjunction of an indeterminate and identification of elements that are congruent with respect to a fixed prime element, are now seen as two modes of construction that lead from rings and fields to other rings and fields respectively.

The elementary axiomatic grounding of geometry also leads to this abstract number concept. Take the case of plane projective geometry. The incidence axioms alone lead to a "number field" that is naturally associated with it. Its elements, the "numbers," are purely geometric entities, namely dilations. A point and a straight line are ratios of triples of "numbers" in that field, $x_1 : x_2 : x_3$ and $u_1 : u_2 : u_3$, respectively, such that incidence of the point $x_1 : x_2 : x_3$ on the line $u_1 : u_2 : u_3$ is represented by the equation

$$x_1 u_1 + x_2 u_2 + x_3 u_3 = 0.$$

Conversely, if one uses these algebraic expressions to define the geometric terms, then every abstract field leads to an associated projective plane that satisfies the incidence axioms. It follows that a restriction involving the number field associated with the projective plane cannot be read off from the incidence axioms. Here the preexisting harmony between geometry and algebra comes to light in the most impressive manner. For the geometric number system to coincide with the continuum of ordinary real numbers, one must introduce axioms of order and continuity, very different in kind from the incidence axioms. We thus arrive at a reversal of the development that has dominated mathematics for centuries and seems to have arisen originally in India and to have been transmitted to the West by Arab scholars: Up till now, we have regarded the number concept as the logical antecedent of geometry, and have therefore approached every realm of magnitudes with a universal and systematically developed number concept independent of the applications involved. Now, however, we revert to the Greek viewpoint that every subject has an associated intrinsic number realm that must be derived from within it. We experience this reversal not only in geometry but also in the new quantum physics. According to quantum physics, the physical magnitudes associated with a particular physical setup (*not* the numerical values that they may take on depending on its different states) admit of an addition and a non-commutative multiplication, and thus give rise to a system of algebraic magnitudes intrinsic to it that cannot be viewed as a sector of the system of real numbers.[3]

And now, as promised, I will present a simple example that illustrates the mutual relation between the topological and abstract-algebraic modes of analysis.

I consider the theory of algebraic functions of a single variable x. Let $K(x)$ be the field of rational functions of x with arbitrary complex coefficients. Let $f(z)$, more precisely $f(z; x)$, be an nth degree polynomial in z with coefficients in $K(x)$. We explained earlier when such a polynomial is said to be irreducible over $K(x)$. This is a purely algebraic concept. Now construct the Riemann surface of the n-valued algebraic function $z(x)$ determined by the equation $f(z; x) = 0$.[4] Its n sheets extend over the x-plane. For easier transformation of the x-plane into the x-sphere by means of a stereographic projection we add to the x-plane a point at infinity. Like the sphere, our Riemann surface is now closed. The irreducibility of the polynomial f is reflected in a very simple topological property of the Riemann surface of $z(x)$, namely its connectedness: if we shake a paper model of that surface it does not break into distinct pieces. Here you witness the coincidence of a purely algebraic and a purely topological concept. Each suggests generalization in a different direction. The algebraic concept of irreducibility depends only on the fact that the coefficients of the polynomial are in a field. In particular, $K(x)$ can be replaced by the field of rational functions of x with coefficients in a preassigned field k which takes the place of the continuum of all complex numbers. On the other hand, from the viewpoint of topology it is irrelevant that the surface in question is a Riemann surface, that it is equipped with a conformal structure, and that it consists of a finite number of sheets that extend over the x-plane. Each of the two antagonists can accuse the other of admitting side issues and of neglecting essential features. Who is right? Questions such as these, involving not facts but ways of looking at facts, can lead to hatred and bloodshed when they touch human emotions. In mathematics, the consequences are not so serious. Nevertheless, the contrast between Riemann's topological theory of algebraic functions and Weierstrass's more algebraically directed school led to a split in the ranks of mathematicians that lasted for almost a generation.

Weierstrass himself wrote to his faithful pupil H. A. Schwarz: "The more I reflect on the principles of function theory — and I do this all the time — the stronger is my conviction that this theory must be established on the foundation of algebraic truths, and that it is therefore not the right way when, contrariwise, the 'transcendent' (to put it briefly) is invoked to establish simple and fundamental algebraic theorems — this is so no matter how attractive are, at a first glance, say, the considerations by means of which Riemann discovered so many of the most important properties of algebraic functions." This strikes us now as one-sided; neither one of the two ways of understanding, the topological or the algebraic, can be acknowledged to have unconditional advantage over the other. And we cannot spare Weierstrass the reproach that he stopped midway. True, he explicitly constructed the functions as algebraic, but he also used as coefficients the algebraically unanalyzed, and in a sense unfathomable for algebraists, continuum of complex numbers. The dominant general theory in the direction followed by Weierstrass is the theory of an abstract number field and its extensions determined by means of algebraic equations. Then the theory of algebraic functions moves in the direction of a shared axiomatic basis with the theory of algebraic numbers. In fact, what suggested to Hilbert his approaches in the theory of number fields was the analogy [between the latter] and the state of things in the realm of algebraic functions discovered by Riemann by his topological methods. (Of course, when it came to proofs, the analogy was useless.)

Our example "irreducible-connected" is typical also in another respect. How visually simple and understandable is the topological criterion (shake the paper model and see if it falls apart) in comparison with the algebraic! The visual primality of the continuum (I think that in this respect it is superior to the 1 and the natural numbers) makes the topological method particularly suitable for both discovery and synopsis in mathematical areas, but is also the cause of difficulties when it comes to rigorous proofs. While it is close to the visual, it is also refractory to logical approaches. That is why Weierstrass, M. Noether and others preferred the laborious, but more solid-feeling, procedure of direct algebraic construction to Riemann's transcendental-topological justification.[5] Now, step by step, abstract algebra tidies up the clumsy computational apparatus. The generality of the assumptions and axiomatization force one to abandon the path of blind computation and to break the complex state of affairs into simple parts that can be handled by means of simple reasoning. Thus algebra turns out to be the El Dorado of axiomatics.

I must add a few words about the method of topology to prevent the picture from becoming altogether vague. If a continuum, say, a two-dimensional closed manifold, a surface, is to be the subject of mathematical investigation, then we must think of it as being subdivided into finitely many "elementary pieces" whose topological nature is that of a circular disk. These pieces are further fragmented by repeated subdivision in accordance with a fixed scheme, and thus a particular spot in the continuum is ever more precisely intercepted by an infinite sequence of nested fragments that arise in the course of successive subdivisions. In the one-dimensional case, the repeated "normal subdivision" of an elementary segment is its bipartition. In the two-dimensional case, each edge is first bipartitioned, then each piece of surface is divided into triangles by means of lines in the surface that lead from an arbitrary center to the (old and new) vertices. What proves that a piece is elementary is that it can be broken into arbitrarily small pieces by repetition of this division process. The scheme of the initial subdivision into elementary pieces — to be referred to briefly in what follows as the "skeleton" — is best described by labeling the surface pieces, edges, and vertices by means of symbols, and thus prescribing the mutual bounding relations of these elements. Following the successive subdivisions, the manifold may be said to be spanned by an increasingly dense net of coordinates which makes it possible to determine a particular point by means of an infinite sequence of symbols that play a role comparable to that of numbers. The reals appear here in the particular form of dyadic fractions, and serve to describe the subdivision of an open one-dimensional continuum. Other than that, we can say that each continuum has its

own arithmetical scheme; the introduction of numerical coordinates by reference
to the special division scheme of an open one-dimensional continuum violates the
nature of things, and its sole justification is the practical one of the extraordinary
convenience of the calculational manipulation of the continuum of numbers with
its four operations. In the case of an actual continuum, the subdivisions can
be realized only with a measure of imprecision; one must imagine that, as the
process of subdivision progresses step by step, the boundaries set by the earlier
subdivisions are ever more sharply fixed. Also, in the case of an actual continuum, the process of subdivision that runs virtually ad infinitum can reach only a
certain definite stage. But in distinction to concrete realization, the localization
in an actual continuum, the combinatorial scheme, the arithmetical nullform, is
a priori determined ad infinitum; and mathematics deals with this combinatorial
scheme alone. Since the continued subdivision of the initial topological skeleton
progresses in accordance with a fixed scheme, it must be possible to read off all
the topological properties of the nascent manifold from that skeleton. This means
that, in principle, it must be possible to pursue topology as finite combinatorics.
For topology, the ultimate elements, the atoms, are, in a sense, the elementary
parts of the skeleton and not the points of the relevant continuous manifold. In
particular, given two such skeletons, it must be possible to decide if they lead
to concurrent manifolds. Put differently, it must be possible to decide if we can
view them as subdivisions of one and the same manifold.

The algebraic counterpart of the transition from the algebraic equation $f(z; x)$
$= 0$ to the Riemann surface is the transition from the latter equation to the
field determined by the function $z(x)$; this is so because the Riemann surface is
uniquely occupied not only by the function $z(x)$ but also by all algebraic functions
in this field. What is characteristic for Riemann's function theory is the converse
problem: given a Riemann surface, construct its field of algebraic functions. The
problem has always just one solution. Since every point $\wp$ of the Riemann surface lies over a definite point of the x-plane, the Riemann surface, as presently
constituted, is embedded in the x-plane. The next step is to abstract from the
embedding relation $\wp \to x$. As a result, the Riemann surface becomes, so to say,
a free-floating surface equipped with a conformal structure and an angle measure. Note that in ordinary surface theory we must learn to distinguish between
the surface as a continuous structure made up of elements of a specific kind, its
points, and the embedding in 3-space that associates with each point $\wp$ of the
surface, in a continuous manner, the point P in space at which $\wp$ is located. In
the case of a Riemann surface, the only difference is that the Riemann surface
and the embedding plane have the same dimension. To abstraction from the
embedding there corresponds, on the algebraic side, the viewpoint of invariance
under arbitrary birational transformations. To enter the realm of topology we
must ignore the conformal structure associated with the free-floating Riemann
surface. Continuing the comparison, we can say that the conformal structure
of the Riemann surface is the equivalent of the metric structure of an ordinary

surface, controlled by the first fundamental form, or of the affine and projective structures associated with surfaces in affine and projective differential geometry respectively.[6] In the continuum of real numbers, it is the algebraic operations of $+$ and $\cdot$ that reflect its structural aspect, and in a continuous group the law that associates with an ordered pair of elements their product plays an analogous role. These comments may have increased our appreciation of the relation of the methods. It is a question of rank, of what is viewed as primary. In topology we begin with the notion of continuous connection, and in the course of specialization we add, step by step, relevant structural features. In algebra this order is, in a sense, reversed. Algebra views the operations as the beginning of all mathematical thinking and admits continuity, or some algebraic surrogate of continuity, at the last step of specialization. The two methods follow opposite directions. Little wonder that they don't get on well together. What is most easily accessible to one is often most deeply hidden to the other. In the last few years, in the theory of representation of continuous groups by means of linear substitutions, I have experienced most poignantly how difficult it is to serve these two masters at the same time. Such classical theories as that of algebraic functions can be made to fit both viewpoints. But viewed from these two viewpoints they present completely different sights.

After all these general remarks I want to use two simple examples that illustrate the different kinds of concept building in algebra and in topology. The classical example of the fruitfulness of the topological method is Riemann's theory of algebraic functions and their integrals. Viewed as a topological surface, a Riemann surface has just one characteristic, namely its connectivity number or genus p. For the sphere $p = 0$ and for the torus $p = 1$. How sensible it is to place topology ahead of function theory follows from the decisive role of the topological number p in function theory on a Riemann surface. I quote a few dazzling theorems: The number of linearly independent everywhere regular differentials on the surface is p. The total order (that is, the difference between the number of zeros and the number of poles) of a differential on the surface is $2p - 2$. If we prescribe more than p arbitrary points on the surface, then there exists just one single-valued function on it that may have simple poles at these points but is otherwise regular; if the number of prescribed poles is exactly p, then, if the points are in general position, this is no longer true. The precise answer to this question is given by the Riemann-Roch theorem in which the Riemann surface enters only through the number p. If we consider all functions on the surface that are everywhere regular except for a single place p at which they have a pole, then its possible orders are all numbers $1, 2, 3, \ldots$ except for certain powers of p (the Weierstrass gap theorem). It is easy to give many more such examples. The genus p permeates the whole theory of functions on a Riemann surface. We encounter it at every step, and its role is direct, without complicated computations, understandable from its topological meaning (provided that we include, once and for all, the Thomson-Dirichlet principle as a fundamental function-theoretic principle).

The Cauchy integral theorem gives topology the first opportunity to enter function theory. The integral of an analytic function over a closed path is 0 only if the domain that contains the path and is also the domain of definition of the analytic function is simply connected. Let me use this example to show how one "topologizes" a function-theoretic state of affairs. If $f(z)$ is analytic, then the integral $\int_\gamma f(z)dz$ associates with every curve a number $F(\gamma)$ such that

(†) $$F(\gamma_1 + \gamma_2) = F(\gamma_1) + F(\gamma_2).$$

$\gamma_1 + \gamma_2$ stands for the curve such that the beginning of γ_2 coincides with the end of γ_1. The functional equation (†) marks the integral $F(\gamma)$ as an additive path function. Also, each point has a neighborhood such that $F(\gamma) = 0$ for each closed path γ in that neighborhood. I will call a path function with these properties a topological integral, or briefly, an integral. In fact, all this concept assumes is that there is given a continuous manifold on which one can draw curves; it is the topological essence of the analytic notion of an integral. Integrals can be added and multiplied by numbers. The topological part of the Cauchy integral theorem states that on a simply connected manifold every integral is homologous to 0 (not only in the small but in the large), that is, $F(\gamma) = 0$ for every closed curve γ on the manifold. In this we can spot the definition of "simply connected." The function-theoretic part states that the integral of an analytic function is a topological integral in our sense of the term. The definition of the order of connectivity [that we are about to state] fits in here quite readily. Integrals $F_1, F_2, \ldots, F_n$ on a closed surface are said to be linearly independent if they are not connected by a homology relation

$$c_1 F_1 + c_2 F_2 + \cdots + c_n F_n \sim 0$$

with constant coefficients c_i other than the trivial one, when all the c_i vanish. The order of connectivity of a surface is the maximal number of linearly independent integrals. For a closed two-sided surface, the order of connectivity h is always an even number $2p$, where p is the genus. From a homology between integrals we can go over to a homology between closed paths. The path homology

$$n_1 \gamma_1 + n_2 \gamma_2 + \cdots + n_r \gamma_r \sim 0$$

states that for every integral F we have the equality

$$n_1 F(\gamma_1) + n_2 F(\gamma_2) + \cdots + n_r F(\gamma_r) = 0.$$

If we go back to the topological skeleton that decomposes the surface into elementary pieces and replace the continuous point-chains of paths by the discrete chains constructed out of elementary pieces, then we obtain an expression for the order of connectivity h in terms of the numbers s, k, and e of pieces, edges, and vertices. The expression in question is the well-known Euler polyhedral formula $h = k - (e + s) + 2$. Conversely, if we start with the topological skeleton, then our reasoning yields the result that this combination h of the number of pieces, edges, and vertices is a topological invariant, namely it has the same value for "equivalent" skeletons which represent the same manifold in different subdivisions.

When it comes to application to function theory, it is possible, using the Thomson-Dirichlet principle, to "realize" the topological integrals as actual integrals of everywhere regular-analytic differentials on a Riemann surface. One can say that all of the constructive work is done on the topological side, and that the topological results are realized in a function-theoretic manner with the help of a universal transfer principle, namely the Dirichlet principle. This is, in a sense, analogous to analytic geometry, where all the constructive work is carried out in the realm of numbers, and then the results are geometrically "realized" with the help of the transfer principle lodged in the coordinate concept.

All this is seen more perfectly in uniformization theory, which plays a central role in all of function theory. But at this point, I prefer to point to another application which is probably close to many of you. I have in mind enumerative geometry, which deals with the determination of the number of points of intersection, singularities, and so on, of algebraic relational structures, which was made into a general, but very poorly justified, system by Schubert and Zeuthen. Here, in the hands of Lefschetz and van der Waerden, topology achieved a decisive success in that it led to definitions of multiplicity valid without exception, as well as to laws likewise valid without exception. Of two curves on a two-sided surface one can cross the other at a point of intersection from left to right or from right to left. These points of intersection must enter every setup with opposite weights $+1$ and -1. Then the total of the weights of the intersections (which can be positive or negative) is invariant under arbitrary continuous deformations of the curves; in fact, it remains unchanged if the curves are replaced by homologous curves. Hence it is possible to master this number through finite combinatorial means of topology and obtain transparent general formulas. Two algebraic curves are, actually, two closed Riemann surfaces embedded in a space of four real dimensions by means of an analytic mapping. But in algebraic geometry a point of intersection is counted with positive multiplicity, whereas in topology one takes into consideration the sense of the crossing. This being so, it is surprising that one can resolve the algebraic question by topological means. The explanation is that in the case of an analytic manifold, crossing always takes place with the same sense. If the two curves are represented in the x_1, x_2-plane in the vicinity of their point of intersection by the functions $x_1 = x_1(s), x_2 = x_2(s)$ and $x_1 = x_1^*(t), x_2 = x_2^*(t)$, then the sense ± 1 with which the first curve intersects the second is given by the sign of the Jacobian

$$\left| \begin{array}{cc} \dfrac{dx_1}{ds} & \dfrac{dx_2}{ds} \\[2ex] \dfrac{dx_1^*}{dt} & \dfrac{dx_2^*}{dt} \end{array} \right| = \frac{\partial(x_1, x_2)}{\partial(s, t)},$$

evaluated at the point of intersection. In the case of complex-algebraic "curves," this criterion always yields the value $+1$. Indeed, let z_1, z_2 be complex coordinates in the plane and let s and t be the respective complex parameters on the two "curves." The real and imaginary parts of z_1 and z_2 play the role of real coordinates in the plane. In their place, we can take $z_1, \bar{z}_1, z_2, \bar{z}_2$. But then the determinant whose sign determines the sense of the crossing is

$$\frac{\partial(z_1, \bar{z}_1, z_2, \bar{z}_2)}{\partial(s, \bar{s}, t, \bar{t})} = \frac{\partial(z_1, z_2)}{\partial(s, t)} \cdot \frac{\partial(\bar{z}_1, \bar{z}_2)}{\partial(\bar{s}, \bar{t})} = \left| \frac{\partial(z_1, z_2)}{\partial(s, t)} \right|^2,$$

and thus invariably positive. Note that the Hurwitz theory of correspondence between algebraic curves can likewise be reduced to a purely topological core.

On the side of abstract algebra, I will emphasize just one fundamental concept, namely the concept of an ideal. If we use the algebraic method, then an algebraic manifold is given in 3-dimensional space with complex Cartesian coordinates x, y, z by means of a number of simultaneous equations

$$f_i(x, y, z) = 0, \ldots, \quad f_n(x, y, z) = 0.$$

The f_i are polynomials. In the case of a curve, it is not at all true that two equations suffice. Not only do the polynomials f_i vanish at points of the manifold but also every polynomial f of the form

$$(**) \qquad f = A_1 f_1 + \cdots + A_n f_n \quad (A_i \text{ are polynomials}).$$

Such polynomials f form an "ideal" in the ring of polynomials. Dedekind defined an ideal in a given ring as a system of ring elements closed under addition and subtraction as well as under multiplication by ring elements. This concept is not too broad for our purposes. The reason is that, according to the Hilbert basis theorem, every ideal in the polynomial ring has a finite basis; there are finitely many polynomials $f_1, \ldots, f_n$ in the ideal such that every polynomial in the ideal can be written in the form (**). Hence the study of algebraic manifolds reduces to the study of ideals. On an algebraic surface there are points and algebraic curves. The latter are represented by ideals that are divisors of the ideal under consideration. The fundamental theorem of M. Noether deals with ideals whose manifold of zeros consists of finitely many points, and makes membership of a polynomial in such an ideal dependent on its behavior at these points. This theorem follows readily from the decomposition of an ideal into prime ideals. The investigations of E. Noether show that the concept of an ideal, first introduced by Dedekind in the theory of algebraic number fields, runs through all of algebra and arithmetic like Ariadne's thread. Van der Waerden was able to justify the enumerative calculus by means of the algebraic resources of ideal theory.

If one operates in an arbitrary abstract number field rather than in the continuum of complex numbers, then the fundamental theorem of algebra, which asserts that every complex polynomial in one variable can be [uniquely] decomposed into linear factors, need not hold. Hence the general prescription in algebraic work: See if a proof makes use of the fundamental theorem or not. In every algebraic theory, there is a more elementary part that is independent of the fundamental theorem, and therefore valid in every field, and a more advanced part for which the fundamental theorem is indispensable. The latter part calls for the algebraic closure of the field. In most cases, the fundamental theorem marks a crucial split; its use should be avoided as long as possible. To establish theorems that hold in an arbitrary field it is often useful to embed the given field in a larger field. In particular, it is possible to embed any field in an algebraically closed field. A well-known example is the proof of the fact that a real polynomial can be decomposed over the reals into linear and quadratic factors. To prove this, we adjoin i to the reals and thus embed the latter in the algebraically closed field of complex numbers. This procedure has an analogue in topology which is used in the study and characterization of manifolds; in the case of a surface, this analogue consists in the use of its covering surfaces.

At the center of today's interest is noncommutative algebra in which one does not insist on the commutativity of multiplication. Its rise is dictated by concrete needs of mathematics. Composition of operations is a kind of noncommutative operation. Here is a specific example. We consider the symmetry properties of functions $f(x_1, x_2, \ldots, x_n)$ of a number of arguments. The latter can be subjected to an arbitrary permutation s. A symmetry property is expressed in one or more equations of the form

$$\sum_s a(s) \cdot sf = 0.$$

Here $a(s)$ stands for the numerical coefficients associated with the permutation. These coefficients belong to a given field K. $\sum_s a(s) \cdot s$ is a "symmetry operator." These operators can be multiplied by numbers, added and multiplied, that is, applied in succession. The result of the latter operation depends on the order of the "factors." Since all formal rules of computation hold for addition and multiplication of symmetry operators, they form a "noncommutative ring" (hypercomplex number system). The dominant role of the concept of an ideal persists in the noncommutative realm. In recent years, the study of groups and their representations by linear substitutions has been almost completely absorbed by the theory of noncommutative rings. Our example shows how the multiplicative group of $n!$ permutations s is extended to the associated ring of magnitudes $\sum_s a(s) \cdot s$ that admit, in addition to multiplication, addition and multiplication by numbers. Quantum physics has given noncommutative algebra a powerful boost.

Unfortunately, I cannot here produce an example of the art of building an abstract-algebraic theory. It consists in setting up the right general concepts, such as fields, ideals, and so on, in decomposing an assertion to be proved into steps (for example, and assertion "A implies B," or $A \to B$, may be decomposed into steps $A \to C, C \to D, D \to B$), and in the appropriate generalization of these partial assertions in terms of general concepts. Once the main assertion has been subdivided in this way and the inessential elements have been set aside, the proofs of the individual steps do not, as a rule, present serious difficulties.

Whenever applicable, the topological method appears, thus far, to be more effective than the algebraic one. Abstract algebra has not yet produced successes comparable to the successes of the topological method in the hands of Riemann. Nor has anyone reached by an algebraic route the peak of uniformization scaled topologically by Klein, Poincaré, and Koebe. Here are questions to be answered in the future. But I do not want to conceal from you the growing feeling among mathematicians that the fruitfulness of the abstracting method is close to exhaustion. It is a fact that beautiful general concepts do not drop out of the sky. The truth is that, to begin with, there are definite concrete problems, with all their undivided complexity, and these must be conquered by individuals relying on brute force. Only then come the axiomatizers and conclude that instead of straining to break in the door and bloodying one's hands, one should have first constructed a magic key of such and such shape and then the door would have opened quietly, as if by itself. But they can construct the key only because the successful breakthrough enables them to study the lock front and back, from the outside and from the inside. Before we can generalize, formalize and axiomatize there must be mathematical substance. I think that the mathematical substance on which we have practiced formalization in the last few decades is near exhaustion and I predict that the next generation will face in mathematics a tough time.[7]

Notes

[A lecture in the summer course of the Swiss Society of Gymnasium Teachers, given in Bern, in October 1931; *WGA* 3:348–358. Translated by Abe Shenitzer, with whose kind permission it is reprinted here, having originally appeared as Weyl 1995 and reprinted in Shenitzer and Stillwell 2002, 149–162.]

[1] [For further explanation of the Dirchlet (or Dirichlet-Thomson) principle, see below, 180.]

[2] [The Tschirnhausen (more often called Tschirnhaus) transformation takes an algebraic equation of the nth degree ($n > 2$) and eliminates its terms of order x^{n-1} and x^{n-2}; see Boyer and Merzbach 1991, 432–433, and Pesic 2003, 66-68.]

[3] [For the history of non-commutativity and its relation to physics, see Pesic 2003, 131–143.]

[4] [Weyl 2009c gives a classic account of Riemann surfaces.]

[5] [M. Noether is Max Noether, an eminent mathematician (1844–1921), the father of Emmy, whose eulogy will be found in the next chapter. For Weyl's commentary on Max, see 49.]

[6] [Weyl 1952a, 77–148, relates metric and affine geometry.]

[7] The sole purpose of this lecture was to give the audience a feeling for the intellectual atmosphere in which a substantial part of modern mathematical research is carried out. For those who wish to penetrate more deeply I give a few bibliographical suggestions. The true pioneers of abstract axiomatic algebra are Dedekind and Kronecker. In our own time, this orientation has been decisively advanced by Steinitz, by E. Noether and her school, and by E. Artin. The first great advance in topology came in the middle of the nineteenth century and was due to Riemann's function theory. The more recent developments are linked primarily to a few works of H. Poincaré devoted to analysis situs (1895–1904). I mention the following books:

1. On algebra: Steinitz, *Algebraic Theory of Fields*, appeared first in *Crelles Journal* in 1910. It was issued as a paperback by R. Baer and H. Hasse and published by Verlag W. de Gruyter, 1930. [Steinitz 1930]

H. Hasse, *Higher Algebra* I, II. Sammlung Goschen 1926/27. [Hasse 1954]

B. van der Waerden, *Modem Algebra* I, II. Springer 1930/31. [Van der Waerden 1970]

2. On topology: H. Weyl, *The Idea of a Riemann Surface*, second ed. Teubner 1923. [Weyl 1964, 2009c]

O. Veblen, *Analysis Situs*, second ed., and S. Lefschetz, *Topology*. Both of these books are in the series *Colloquium Publications of the American Mathematical Society*, New York 1931 and 1930 respectively. [Veblen 1931, Lefschetz 1956]

3. Volume I of F. Klein, *History of Mathematics in the Nineteenth Century*, Springer 1926. [Klein 1979]

Emmy Noether
(1935)

With deep dismay Emmy Noether's friends living in America learned about her sudden passing away on Sunday, April 14. She seemed to have got well over an operation for tumor; we thought her to be on the way to convalescence when an unexpected complication led her suddenly on the downward path to her death within a few hours. She was such a paragon of vitality, she stood on the earth so firm and healthy with a certain sturdy humor and courage for life, that nobody was prepared for this eventuality. She was at the summit of her mathematical creative power; her far-reaching imagination and her technical abilities accumulated by continued experience, had come to a perfect balance; she had eagerly set to work on new problems. And now suddenly — the end, her voice silenced, her work abruptly broken off.

> Down, down, down into the darkness of the grave
> Gently they go, the beautiful, the tender, the kind;
> Quietly they go, the intelligent, the witty, the brave.
> I know. But I do not approve. And I am not resigned.

A mood of defiance similar to that expressed in this "Dirge without music" by Edna St. Vincent Millay, mingles with our mourning in the present hour when we are gathered to commemorate our friend, her life and work and personality.

I am not able to tell much about the outward story of her life; far from her home and those places where she lived and worked in the continuity of decades, the necessary information could not be secured. She was born March 23, 1882, in the small South German university town of Erlangen. Her father was Max Noether, himself a great mathematician who played an important role in the development of the theory of algebraic functions as the chief representative of the algebraic-geometric school. He had come to the University of Erlangen as a professor of mathematics in 1875, and stayed there until his death in 1921. Besides Emmy there grew up in the house her brother Fritz, younger by two and a half years. He turned to applied mathematics in later years, was until recently professor at the Technische Hochschule in Breslau, and by the same fate that ended Emmy's career in Göttingen is now driven off to the Research Institute for

Mathematics and Mechanics in Tomsk, Siberia. The Noether family is a striking example of the hereditary nature of the mathematical talent, the most shining illustration of which is the Basle Huguenot dynasty of the Bernoullis.

Side by side with Noether acted in Erlangen as a mathematician the closely befriended Gordan, an offspring of Clebsch's school like Noether himself. Gordan had come to Erlangen shortly before, in 1874, and he, too, remained associated with that university until his death in 1912. Emmy wrote her doctor's thesis under him in 1907: "On complete systems of invariants for ternary biquadratic forms"; it is entirely in line with the Gordan spirit and his problems. The *Mathematische Annalen* contains a detailed obituary of Gordan and an analysis of his work, written by Max Noether with Emmy's collaboration. Besides her father, Gordan must have been well-nigh one of the most familiar figures in Emmy's early life, first as a friend of the house, later as a mathematician also; she kept a profound reverence for him though her own mathematical taste soon developed in quite a different direction. I remember that his picture decorated the wall of her study in Göttingen. These two men, the father and Gordan, determined the atmosphere in which she grew up. Therefore I shall venture to describe them with a few strokes.

Riemann had developed the theory of algebraic functions of one variable and their integrals, the so-called Abelian integrals, by a function-theoretic transcendental method resting on the minimum principle of potential theory which he named after Dirichlet, and had uncovered the purely topological foundations of the manifold function-theoretic relations governing this domain. (Stringent proof of Dirichlet's principle which seemed so evident from the physicist's standpoint was only given about fifty years later by Hilbert.)[1] There remained the task of replacing and securing his transcendental existential proofs by the explicit algebraic construction starting with the equation of the algebraic curve. Weierstrass solved this problem (in his lectures published in detail only later) in his own half function-theoretic, half algebraic way, but Clebsch had introduced Riemann's ideas into the geometric theory of algebraic curves and Noether became, after Clebsch had passed away young, his executor in this matter: he succeeded in erecting the whole structure of the algebraic geometry of curves on the basis of the so-called Noether residual theorem. This line of research was taken up later on, mainly in Italy; the vein Noether struck is still a profusely gushing spring of investigations; among us, men like Lefschetz and Zariski bear witness thereto. Later on there arose, beside Riemann's transcendental and Noether's algebraic-geometric method, an arithmetical theory of algebraic functions due to Dedekind and Weber on the one side, to Hensel and Landsberg on the other. Emmy Noether stood closer to this trend of thought. A brief report on the arithmetical theory of algebraic functions that parallels the corresponding notions in the competing theories was published by her in 1920 in the *Jahresbenchte der Deutschen Mathematikervereinigung*. She thus supplemented the well-known report by Brill and her father on the algebraic-geometric theory that had appeared in 1894 in one of the first volumes of the *Jahresberichte*. Noether's residual theorem was later fitted by Emmy into her general theory of ideals in arbitrary rings. This scientific kinship of father and daughter — who became in a certain sense his successor in algebra, but stands beside him independent in her fundamental attitude and in her problems — is something extremely beautiful and gratifying. The father was —such is the impression I gather from his papers and even more from the many obituary biographies he wrote for the *Mathematische Annalen* — a very intelligent, warm-hearted harmonious man of many-sided interests and sterling education.

Gordan was of a different stamp. A queer fellow, impulsive and one-sided. A great walker and talker — he liked that kind of walk to which frequent stops at a beer-garden or a café belong. Either with friends, and then accompanying his discussions with violent gesticulations, completely irrespective of his surroundings; or alone, and then murmuring to himself and pondering over mathematical problems; or if in an idler mood, carrying out long numerical calculations by heart. There always remained something of the eternal "Bursche" of the 1848 type about him — an air of dressing gown, beer and tobacco, relieved however by a keen sense of humor and a strong dash of wit.[2] When he had to listen to others, in classrooms or at meetings, he was always half asleep. As a mathematician not of Noether's rank, and of an essentially different kind. Noether himself concludes his characterization of him with the short sentence: "Er war ein Algorithmiker." ["He was a calculator."] His strength rested on the invention and calculative execution of formal processes. There exist papers of his where twenty pages of formulas are not interrupted by a single text word; it is told that in all his papers he himself wrote the formulas only, the text being added by his friends. Noether says of him: "The formula always and everywhere was the indispensable support for the formation of his thoughts, his conclusions and his mode of expression.... In his lectures he carefully avoided any fundamental definition of conceptual kind, even that of the limit."

He, too, had belonged to Clebsch's most intimate collaborators, had written with Clebsch their book on Abelian integrals; he later shifted over to the theory of invariants following his formal talent; here he added considerably to the development of the so-called symbolic method, and he finally succeeded in proving by means of this computative method of explicit construction the finiteness of a rational integral basis for binary invariants. Years later Hilbert demonstrated the theorem much more generally for an arbitrary number of variables — by an entirely new approach, the characteristic Hilbertian species of methods, putting aside the whole apparatus of symbolic treatment and attacking the thing itself as directly as possible. *Ex ungue leonem* — the young lion Hilbert showed his claws.[3] It was, however, at first only an existential proof providing for no actual finite algebraic construction. Hence Gordan's characteristic exclamation: "This is not mathematics, but theology!" What then would he have said about his former pupil Emmy Noether's later "theology," that abhorred all calculation and operated in a much thinner air of abstraction than Hilbert ever dared!

Gordan once struck upon a formal analogy between binary invariants and the scheme of valence bonds in chemistry — the same analogy by which Sylvester had been surprised many years before when thinking about an illustration of invariant theory appropriate for an audience of laymen; it is the subject of Sylvester's paper in the first volume of the *American Journal of Mathematics* founded by him at Johns Hopkins. Gordan seems to have been unaware of his predecessor. Anyway, he was led by his little discovery to propose the establishment of chairs for a new science, "mathematical chemistry," all over the German universities; I mention this as an incident showing his impetuosity and lack of survey. By the way, modern quantum mechanics recently has changed this analogy into a true theory disclosing the binary invariants as the mathematical tool for describing the several valence states of a molecule in spin space.[4]

The meteor Felix Klein, whose mathematical genius caught fire through the collision of Riemann's and Galois' worlds of ideas, skimmed Erlangen before Emmy was born; he promulgated there his "Erlanger Programm," but soon moved on to Munich.[5] By him Gordan was inspired to those invariant theoretical investigations that center around Klein's book on the icosahedron and the adjoint questions in the theory of algebraic equations.[6] Even after their local separation both continued in their intense cooperation — a queer contrasting team if one comes to think of Gordan's formal type and Klein's, entirely oriented by intuition. The general problem at the bottom of their endeavors, Klein's form problem has likewise stayed alive to our days and quite recently has undergone a new deep-reaching treatment by Dr. Brauer's applying to it the methods of hypercomplex number systems and their representations which formed the main field of Emmy Noether's activities during the last six or seven years.[7]

It is queer enough that a formalist like Gordan was the mathematician from whom her mathematical orbit set out; a greater contrast is hardly imaginable than between her first paper, the dissertation, and her works of maturity; for the former is an extreme example of formal computations and the latter constitute an extreme and grandiose example of conceptual axiomatic thinking in mathematics. Her thesis ends with a table of the complete system of covariant forms for a given ternary quartic consisting of not less than 331 forms in symbolic representation. It is an awe-inspiring piece of work; but today I am afraid we should be inclined to rank it among those achievements with regard to which Gordan himself once said when asked about the use of the theory of invariants: "Oh, it is very useful indeed; one can write many theses about it."

It is not quite easy to evoke before an American audience a true picture of that state of German life in which Emmy Noether grew up in Erlangen; maybe the present generation in Germany is still more remote from it. The great stability of burgher life was in her case accentuated by the fact that Noether (and Gordan too) were settled at one university for so long an uninterrupted period. One may dare to add that the time of the primary proper impulses of their production was gone, though they undoubtedly continued to be productive mathematicians; in this regard, too, the atmosphere around her was certainly tinged by a quiet uniformity. Moreover, there belongs to the picture the high standing, and the great solidity in the recognition of, spiritual values; based on a solid education, a deep and genuine active interest in the higher achievements of intellectual culture, and on a well-developed faculty of enjoying them. There must have prevailed in the Noether home a particularly warm and companionable family life. Emmy Noether herself was, if I may say so, warm like a loaf of bread. There irradiated from her a broad, comforting, vital warmth. Our generation accuses that time of lacking all moral sincerity, of hiding behind its comfort and bourgeois peacefulness, and of ignoring the profound creative and terrible forces that really shape man's destiny; moreover of shutting its eyes to the contrast between the spirit of true Christianity which was confessed, and the private and public life as it was actually lived. Nietzsche arose in Germany as a great awakener. It is hardly possible to exaggerate the significance which Nietzsche (whom by the way Noether once met in the Engadin) had in Germany for the thorough change in the moral and mental atmosphere. I think he was fundamentally right — and yet one should not deny that in wide circles in Germany, as with the Noethers, the esteem in which the spiritual goods were held, the intellectual culture, good-heartedness, and human warmth were thoroughly genuine — notwithstanding their sentimentality, their Wagnerianism, and their plush sofas.

Emmy Noether took part in the housework as a young girl, dusted and cooked, and went to dances, and it seems her life would have been that of an ordinary woman had it not happened that just about that time it became possible in Germany for a girl to enter on a scientific career without meeting any too marked resistance. There was nothing rebellious in her nature; she was willing to accept conditions as they were. But now she became a mathematician. Her dependence on Gordan did not last long; he was important as a starting point, but was not of lasting scientific influence upon her. Nevertheless the Erlangen mathematical air may have been responsible for making her into an algebraist. Gordan retired in 1910; he was followed first by Erhard Schmidt, and the next year by Ernst Fischer. Fischer's field was algebra again, in particular the theory of elimination and of invariants. He exerted upon Emmy Noether, I believe, a more penetrating influence than Gordan did. Under his direction the transition from Gordan's formal standpoint to the Hilbert method of approach was accomplished. She refers in her papers at this time again and again to conversations with Fischer. This epoch extends until about 1919. The main interest is concentrated on finite rational and integral bases; the proof of finiteness is given by her for the invariants of a finite group (without using Hilbert's general basis theorem for ideals), for invariants with restriction to integral coefficients, and finally she attacks the same question along with the question of a minimum basis consisting of independent elements for fields of rational functions.[8]

Already in Erlangen about 1913 Emmy lectured occasionally, substituting for her father when he was taken ill. She must have been to Göttingen about that time, too, but I suppose only on a visit with her brother Fritz. At least I remember him much better than her from my time as a Göttinger *Privatdozent*, 1910–1913. During the war, in 1916, Emmy came to Göttingen for good; it was due to Hilbert's and Klein's direct influence that she stayed. Hilbert at that time was over head and ears in the general theory of relativity, and for Klein, too, the theory of relativity and its connection with his old ideas of the Erlangen program brought the last flareup of his mathematical interests and mathematical production. The second volume of his history of mathematics in the nineteenth century bears witness thereof. To both Hilbert and Klein Emmy was welcome as she was able to help them with her invariant-theoretic knowledge. For two of the most significant sides of the general relativity theory she gave at that time the genuine and universal mathematical formulation: First, the reduction of the problem of differential invariants to a purely algebraic one by use of "normal coordinates"; second, the identities between the left sides of Euler's equations of a problem of variation which occur when the (multiple) integral is invariant with respect to a group of transformations involving arbitrary functions (identities that contain the conservation theorem of energy and momentum in the case of invariance with respect to arbitrary transformations of the four world coordinates).[9]

Still during the war, Hilbert tried to push through Emmy Noether's "Habilitation" in the Philosophical Faculty in Göttingen. He failed due to the resistance of the philologists and historians. It is a well-known anecdote that Hilbert supported her application by declaring at the faculty meeting, "I do not see that the sex of the candidate is an argument against her admission as *Privatdozent*. After all, we are a university and not a bathing establishment." Probably he provoked the adversaries even more by that remark. Nevertheless, she was able to give lectures in Göttingen, that were announced under Hilbert's name. But in 1919, after the end of the War and the proclamation of the German Republic had changed the conditions, her Habilitation became possible. In 1922 there followed her nomination as a *"nichtbeamteter ausserordentlicher Professor"*; this was a mere title carrying no obligations and no salary. She was, however, entrusted with a *"Lehrauftrag"* for algebra, which carried a modest remuneration.[10]

During the wild times after the Revolution of 1918, she did not keep aloof from the political excitement, she sided more or less with the Social Democrats; without being actually in party life she participated intensely in the discussion of the political and social problems of the day. One of her first pupils, Grete Hermann, belonged to Nelson's philosophic-political circle in Göttingen. It is hardly imaginable nowadays how willing the young generation in Germany was at that time for a fresh start, to try to build up Germany, Europe, society in general, on the foundations of reason, humaneness, and justice. But alas! the mood among the academic youth soon enough veered around; in the struggles

that shook Germany during the following years and which took on the form of
civil war here and there, we find them mostly on the side of the reactionary
and nationalistic forces. Responsible for this above all was the breaking by the
Allies of the promise of Wilson's "Fourteen Points," and the fact that Republican
Germany came to feel the victors' fist not less hard than the Imperial Reich could
have; in particular, the youth were embittered by the national defamation added
to the enforcement of a grim peace treaty. It was then that the great opportunity
for the pacification of Europe was lost, and the seed sown for the disastrous
development we are the witnesses of. In later years Emmy Noether took no part
in matters political. She always remained, however, a convinced pacifist, a stand
which she held very important and serious.

In the modest position of a *"nicht-beamteter ausserordentlicher Professor"*
she worked in Göttingen until 1933, during the last years in the beautiful new
Mathematical Institute that had risen in Göttingen chiefly by Courant's energy
and the generous financial help of the Rockefeller Foundation. I have a vivid
recollection of her when I was in Göttingen as visiting professor in the winter
semester of 1926–1927, and lectured on representations of continuous groups.
She was in the audience; for just at that time the hypercomplex number systems and their representations had caught her interest and I remember many
discussions when I walked home after the lectures, with her and von Neumann,
who was in Göttingen as a Rockefeller Fellow, through the cold, dirty, rain-wet
streets of Göttingen. When I was called permanently to Göttingen in 1930, I
earnestly tried to obtain from the Ministerium a better position for her, because
I was ashamed to occupy such a preferred position beside her whom I knew to be
my superior as a mathematician in many respects. I did not succeed, nor did an
attempt to push through her election as a member of the Göttinger Gesellschaft
der Wissenschaften. Tradition, prejudice, external considerations, weighted the
balance against her scientific merits and scientific greatness, by that time denied by no one. In my Göttingen years, 1930–1933, she was without doubt the
strongest center of mathematical activity there, considering both the fertility of
her scientific research program and her influence upon a large circle of pupils.

Her development into that great independent master whom we admire today
was relatively slow. Such a late maturing is a rare phenomenon in mathematics; in most cases the great creative impulses lie in early youth. Sophus Lie,
like Emmy Noether, is one of the few great exceptions. Not until 1920, thirteen
years after her promotion, appeared in the *Mathematische Zeitschrift* that paper
of hers written with Schmeidler, "Über Moduln in nicht-kommutativen Bere-
ichen, insbesondere aus Differential- und Differenzen-Ausdrücken" ["On Modules
in Non-Commutative Fields, Especially from Differential and Difference Expressions"], which seems to mark the decisive turning point. It is here for the first
time that the Emmy Noether appears whom we all know, and who changed the
face of algebra by her work. Above all, her conceptual axiomatic way of thinking
in algebra becomes first noticeable in this paper dealing with differential opera-

tors as they are quite common nowadays in quantum mechanics. In performing
them, one after the other, their composition, which may be interpreted as a kind
of multiplication, is not commutative. But instead of operating with the formal
expressions, the simple properties of the operations of addition and multiplication to which they lend themselves are formulated as axioms at the beginning of
the investigation, and these axioms then form the basis of all further reasoning.
A similar procedure has remained typical for Emmy Noether from then on. Later
I shall try to characterize this world of algebra as a whole in which the scene of
her mathematical activities was laid.

Not less characteristic for Emmy was her collaboration with another, in this
case with Schmeidler. I suppose that Schmeidler gave as much as he received in
this cooperation. In later years, however, Emmy Noether frequently acted as the
true originator; she was most generous in sharing her ideas with others. She had
many pupils, and one of the chief methods of her research was to expound her
ideas in a still unfinished state in lectures, and then discuss them with her pupils.
Sometimes she lectured on the same subject one semester after another, the whole
subject taking on a better ordered and more unified shape every time, and gaining
of course in the substance of results. It is obvious that this method sometimes
put enormous demands upon her audience. In general, her lecturing was certainly
not good in technical respects. For that she was too erratic and she cared too
little for a nice and well arranged form. And yet she was an inspired teacher; he
who was capable of adjusting himself entirely to her, could learn very much from
her. Her significance for algebra cannot be read entirely from her own papers;
she had great stimulating power and many of her suggestions took final shape
only in the works of her pupils or co-workers. A large part of what is contained in
the second volume of van der Waerden's "Modern Algebra" must be considered
her property. The same is true of parts of Deuring's recently published book on
algebras in which she collaborated intensively. Hasse acknowledges that he owed
the suggestion for his beautiful papers on the connection between hypercomplex
quantities and the theory of class fields to casual remarks by Emmy Noether.
She could just utter a far-seeing remark like this, "Norm rest symbol is nothing
else than cyclic algebra" in her prophetic lapidary manner, out of her mighty
imagination that hit the mark most of the time and gained in strength in the
course of years; and such a remark could then become a signpost to point the way
for difficult future work. And one cannot read the scope of her accomplishments
from the individual results of her papers alone: she originated above all a new
and epoch-making style of thinking in algebra.

She lived in close communion with her pupils; she loved them, and took interest in their personal affairs. They farmed a somewhat noisy and stormy family,
"the Noether boys" as we called them in Göttingen. Among her pupils proper
I may name Grete Hermann, Krull, Holzer, Grell, Koethe, Deuring, Fitting,
Witt, Tsen, Shoda, Levitzki. F. K. Schmidt is strongly influenced by her, chiefly
through Krull's mediation. Van der Waerden came to her from Holland as a more

or less finished mathematician and with ideas of his own; but he learned from Emmy Noether the apparatus of notions and the kind of thinking that permitted him to formulate his ideas and to salve his problems. Artin and Hasse stand beside her as two independent minds whose field of production touches on hers closely, though both have a stronger arithmetical texture. With Hasse above all she collaborated very closely during her last years. From different sides, Richard Brauer and she dealt with the profounder structural problems of algebras, she in a more abstract spirit, Brauer, educated in the school of the great algebraist I. Schur, more concretely operating with matrices and representations of groups; this, too, led to an extremely fertile cooperation. She held a rather close friendship with Alexandroff in Moscow, who came frequently as a guest to Göttingen. I believe that her mode of thinking has not been without influence upon Alexandroff's topological investigations.[11] About 1930 she spent a semester in Moscow and there got into close touch with Pontrjagin also. Before that, in 1928–1929, she had lectured for one semester in Frankfurt while Siegel delivered a course of lectures as a visitor in Göttingen.

In the spring of 1933 the storm of the National Revolution broke over Germany. The Göttinger Mathematisch-Naturwissenschaftliche Fakultät, for the building up and consolidation of which Klein and Hilbert had worked for decades, was struck at its roots. After an interregnum of one day by Neugebauer, I had to take over the direction of the Mathematical Institute. But Emmy Noether, as well as many others, was prohibited from participation in all academic activities, and finally her *venia legendi*, as well as her "*Lehrauftrag*" and the salary going with it, were withdrawn.[12] A stormy time of struggle like this one we spent in Göttingen in the summer of 1933 draws people closer together; thus I have a particularly vivid recollection of these months. Emmy Noether, her courage, her frankness, her unconcern about her own fate, her conciliatory spirit, were, in the midst of all the hatred and meanness, despair and sorrow surrounding us, a moral solace. It was attempted, of course, to influence the Ministerium and other responsible and irresponsible but powerful bodies so that her position might be saved. I suppose there could hardly have been in any other case such a pile of enthusiastic testimonials filed with the Ministerium as was sent in on her behalf. At that time we really fought; there was still hope left that the worst could be warded off. It was in vain. Franck, Born, Courant, Landau, Emmy Noether, Neugebauer, Bernays and others — scholars the university had before been proud of — had to go because the possibility of working was taken away from them. Göttingen scattered into the four winds! This fate brought Emmy Noether to Bryn Mawr, and the short time she taught here and as guest at our Institute for Advanced Study in Princeton is still too fresh in our memory to need to be spoken of. She harbored no grudge against Göttingen and her fatherland for what they had done to her. She broke no friendship on account of political dissension. Even last summer she returned to Göttingen, and lived and worked there as though all things were as before. She was sincerely glad that Hasse was endeavoring with success to rebuild the old, honorable and proud mathematical tradition of Göttingen even in the changed political circumstances. But she had adjusted herself with perfect ease to her new American surroundings, and her girl students here were as near to her heart as the Noether boys had been in Göttingen. She was happy at Bryn Mawr; and indeed perhaps never before in her life had she received so many signs of respect, sympathy, friendship, as were bestowed upon her during her last one and a half years at Bryn Mawr. Now we stand at her grave.

It shall not be forgotten what America did during these last two stressful years for Emmy Noether and for German science in general.

If this sketch of her life is to be followed by a short synopsis of her work and her human and scientific personality, I must attempt to draw in a few strokes the scene of her work: the world of algebra. The system of real numbers, of so paramount import for the whole of mathematics and physics, resembles a Janus head with two faces: In one aspect it is the field of the algebraic operations + and ×, and their inversions. In the other aspect it is a continuous manifold, the parts of which are continuously connected with each other. The one is the algebraic, the other the topological face of numbers. Modern axiomatics, single-minded as it is and hence disliking this strange mixture of war and peace (in this respect differing from modern politics), carefully disjointed both parts.

Hence the pure algebraist can do nothing with his numbers except perform upon them the four species, addition, subtraction, multiplication, and division. For him, therefore, a set of numbers is closed, he has no means to get beyond it when these operations applied to any two numbers of the set always lead to a number of the same set again. Such a set is called a domain of rationality or a field. The simplest field is the set of all rational numbers. Another example is the set of the numbers of the form $a + b\sqrt{2}$ where a and b are rational, the so-called algebraic number field $(\sqrt{2})$. The classical problem of algebra is the solution of an algebraic equation $f(x) = 0$ whose coefficients may lie in a field K, for instance the field of rational numbers. Knowing a root δ of the equation, one knows at the same time all numbers arising from δ (and the numbers of K) by means of the four species: they form the algebraic field $K(\delta)$ comprising K. Within this number field $K(\delta)$, δ itself plays the role of a determining number from which all other numbers can be rationally derived. But many, almost all, numbers of $K(\delta)$ can take the place of δ in this respect. It is, therefore, a great advance to replace the study of the equation $f(x) = 0$ by the study of the field $K(\delta)$. We thereby extinguish unessential features, we take uniformly into account all equations arising from the one $f(x) = 0$ by rational transformations of the unknown x, and we replace a formula, the equation $f(x) = 0$, which might seduce us to blind computations, by a notion, the notion of the field which one can get at only in a conceptual way.

Within the system of *integral* numbers the operations of addition, subtraction, and multiplication only allow unlimited performance; division has to be canceled.

Such a domain is called a domain of integrity or a *ring*. As the notion of integer is characteristic of number theory, one may say: number theory deals with rings instead of fields. The polynomials of one variable or indeterminate x are likewise such a domain of quantities as we described to form a ring; the coefficients of the polynomials might here be restricted to a given number field or ring. Algebra does not interpret the argument x to be a variable varying over a continuous range of values; it looks upon it as an indeterminate, an empty symbol serving only to weld the coefficients of the polynomial into a unified expression which suggests in a natural way the rules of addition and multiplication. The statement that a polynomial vanishes means that all its coefficients are zero rather than that the function takes on the value zero for all values of the independent variable. One is not forbidden to substitute an indeterminate x by a number or by a polynomial of one or several other indeterminates $y, z, \ldots$; however, this is a formal process projecting the ring of polynomials of x faithfully upon the ring of numbers or of polynomials in $y, z, \ldots$. Faithfully, that means preserving all rational relations expressible in terms of the fundamental operations, addition, subtraction, multiplication.

Besides adjunction of indeterminates, algebra knows another procedure for forming new fields or rings. Let p be a prime number, for instance 5. We take the ordinary integers, agreeing, however, to consider numbers to be equal when they are congruent modulo p, i.e., when they give the same remainder under division by p. One may illustrate this by winding the line of numbers on a circle of circumference p. A peculiar field then arises consisting of p different elements only. To the *prime number* there corresponds within the ring of polynomials of a single variable x (with numerical coefficients taken from a given number field K) the *prime polynomial* $p(x)$. By considering two polynomials equal which are congruent modulo a given prime polynomial $p(x)$, the ring of all polynomials is changed into a *field* which possesses exactly the same algebraic properties as the number field $K(\delta)$ arising from the underlying number field K by adjoining a root δ of the equation $p(x) = 0$. But the present process goes on within pure algebra without requiring solution of an equation $p(x) = 0$ that is actually unsolvable in K. This interpretation of the algebraic number fields $K(\delta)$ was given by Kronecker after Cauchy had already founded the calculation with the imaginary number i on this idea.

In such a way one was led by degrees to erect algebra in a purely axiomatic manner. A whole array of great mathematical names could be mentioned who initiated and developed this axiomatic trend: after Kronecker and Dedekind, E. H. Moore in America, Peano in Italy, Steinitz, and, above all, Hilbert in Germany. A field now is a realm of elements, called numbers, within which two operations $+$ and $\times$ are defined, satisfying the usual axioms. If one leaves out the axiom of division which states the unique invertibility of multiplication, one gets a ring instead of a field. The fields no longer appear as parts cut out of that universal realm of numbers, the continuum of the real or complex numbers that the Calculus is concerned with, but every field is now, so to speak, a world in itself. One may join the elements of any field by operations, but not the elements of different fields. This standpoint that each object which is offered to mathematical analysis carries its own kind of numbers to be defined in terms of that object and its intrinsic constituents, instead of approaching every object by the same universal number system developed a priori and independently of the applications — this standpoint, I say, has gained ground more and more also in the axiomatic foundations of geometry and recently in a rather surprising manner in quantum physics. We are here confronted by one of those mysterious parallelisms in the development of mathematics and physics that might induce one to believe in a pre-established harmony between nature and mind.

When speaking of axiomatics, I was referring to the following methodical procedure: One separates in a natural way the different sides of a concretely given object of mathematical investigation, makes each of them accessible from its own relatively narrow and easily surveyable group of assumptions, and then by joining the partial results after appropriate specialization, returns to the complex whole. The last synthetic part is purely mechanical. The art lies in the first analytical part of breaking up the whole and generalizing the parts. One does not seek the general for the sake of generality, but the point is that each generalization simplifies by reducing the hypotheses and thus makes us understand certain sides of an unsurveyable whole. Whether a partition with corresponding generalization is natural, can hardly be judged by any other criterion than its fertility. If one systematizes this procedure which the individual investigator manages supported by all the analogies available to him by the mass of his mathematical experiences and with more or less inventive ability and sensitivity, one comes upon axiomatics. Hence axiomatics is today by no means merely a method for logical clarification and deepening of the foundations, but it has become a powerful weapon of concrete mathematical research itself. This method was applied by Emmy Noether with masterly skill, it suited her nature, and she made algebra the Eldorado of axiomatics. An important point is the ascertainment of the "right" general notions like field, ring, ideal, etc., the splitting-up of a proposition into partial propositions and their right generalizations by means of those general notions. This partition of the whole and screening off of the unessential features once accomplished, the proof of the individual steps does not cause any serious trouble in many cases. In a conference on topology and abstract algebra as two ways of mathematical understanding, in 1931, I said this:

> Nevertheless I should not pass over in silence the fact that today the feeling among mathematicians is beginning to spread that the fertility of these abstracting methods is approaching exhaustion. The case is this; that all these nice general notions do not fall into our laps by themselves. But definite concrete problems were first conquered in their undivided complexity, single-handed by brute force, so to speak. Only afterwards the axiomaticians came along and stated: Instead of

breaking in the door with all your might and bruising your hands, you should have constructed such and such a key of skill, and by it you would have been able to open the door quite smoothly. But they can construct the key only because they are able, after the breaking in was successful, to study the lock from within and without. Before you can generalize, formalize and axiomatize, there must be a mathematical substance. I think that the mathematical substance in the formalizing of which we have trained ourselves during the last decades, becomes gradually exhausted. And so I foresee that the generation now rising will have a hard time in mathematics.[13]

Emmy Noether protested against that: and indeed she could point to the fact that just during the last years the axiomatic method had disclosed in her hands new, concrete, profound problems by the application of non-commutative algebra upon commutative fields and their number theory, and had shown the way to their solution.

Emmy Noether's scientific production seems to me to fall into three clearly distinct epochs: (1) the period of relative dependence, 1907–1919; (2) the investigations grouped around the general theory of ideals, 1920–26; (3) the study of the non-commutative algebras, their representations by linear transformations, and their application to the study of commutative number fields and their arithmetics, from 1927 on. The first epoch was described in the sketch of her life. I should now like to say a few words about the second epoch, the epoch of the general theory of ideals.

The ideals had been devised by Dedekind in order to reestablish, by introducing appropriate ideal elements, the main law of unique decomposition of a number into prime factors that broke down in algebraic number fields. The thought consisted in replacing a number, like 6 for instance, in its property as a divisor by the set of all numbers divisible by 6; this set is called the ideal (6). In the same manner one may interpret the greatest common divisor of two numbers, a, b, as the set of all numbers of form $ax + by$ where x, y range independently over all integers. In the ring of ordinary integers this system is identical with a system of the multiples of a single number d, the greatest common divisor. This, however, is not the case in algebraic number fields, and hence it becomes necessary to admit as divisors not only numbers but also ideals. An ideal in a ring R then has to be defined as a subset of R such that sum and difference of two numbers of the ideal belong to the ideal as well as the product of a number of the ideal by an arbitrary number of the ring. Still, from another side, this notion appeared in algebraic geometry. An algebraic surface in space is defined by one algebraic equation $f = 0$; here f is a polynomial with respect to the coordinates. If one is to consider algebraic manifolds of fewer dimensions, one has to put down instead a finite system of algebraic equations $f_1 = 0, f_2 = 0, \ldots, f_h = 0$. But then all polynomials vanish upon the algebraic manifold which arise by linear combination of the basic polynomials $f_1, f_2, \ldots, f_h$ in the form $A_1 f_1 + A_2 f_2 + \cdots + A_h f_h$ where the *A*s are quite arbitrary polynomials. All the polynomials of this kind form an ideal in the ring of polynomials; the algebraic manifold consists of the points in which all polynomials of the ideal vanish. With such ideals Hilbert's basis theorem was concerned, one of the chief tools in Hilbert's study of invariants; it asserts that every ideal of polynomials has a finite basis. Noether's residual theorem contains a criterion that allows us to decide whether a polynomial belongs to an ideal the members of which have in common only a finite number of zeros. For ideals of polynomials Lasker — better known to non-mathematicians as world chess champion for many years — obtained results which showed that their laws depart considerably from those met by Dedekind in the algebraic number fields.

Consider, for instance, the following three rings: the ring of ordinary integers, the rings of polynomials of one and of two independent variables with rational coefficients. The theorem of unique decomposition into prime factors holds in each of them; but Euclid's algorithm or the fact that the greatest common divisor of two elements, a, b, is contained in the ideal (a, b), i.e., can be expressed in the form $af + bg$ by means of two appropriate elements, f, g, of the ring, is true only in the first two cases. Indeed, in the domain of polynomials of two indeterminates x and y, the polynomials x and y themselves have no common divisor; nevertheless an equation like $1 = xf + yg$ where f and g are two polynomials, is impossible as the right side vanishes at the origin $x = 0, y = 0$.

Emmy Noether developed a general theory of ideals on an axiomatic basis that comprised all cases. Her chief axiom is the *Teilerkettensatz*: the hypothesis that a chain of ideals $\mathfrak{a}_1, \mathfrak{a}_2, \mathfrak{a}_3, \ldots$ necessarily comes to an end after a finite number of steps if each term $\mathfrak{a}_i$ comprises the preceding $\mathfrak{a}_{i-1}$ as a proper part.[14] By her abstract theory many important developments of mathematics are welded together. Moreover, she showed how one can descend in the same axiomatic manner to the polynomial ideals on the one hand, and to the classical case of ideals in algebraic number fields on the other hand. In some instances her general theory passes even beyond what was known before through Lasker for polynomial ideals.

Until now we have stuck to all axioms satisfied by the ordinary numbers. There exist, however, strong motives for abandoning the commutative law of multiplication. Indeed, operations like the rotations of a rigid body in space, are entities which behave with respect to their composition in a non-commutative fashion: for the composition of two rotations it really matters whether one first performs the first and then the second, or does it in inverse order. Composition is here considered as a kind of multiplication. Rotations when expressed in terms of coordinates are linear transformations. The linear transformations, as they are capable of addition and composition or multiplication, form the most important example of non-commutative quantities. One therefore attempts to realize any given abstract non-commutative ring or "algebra" of quantities by linear transformations without destroying the relations established among them by the fundamental operations $+$ and $\times$; this is the aim of the theory of representations.

The theory of non-commutative algebras and their representations was built up by Emmy Noether in a new unified, purely conceptual manner by making use of all results that had been accumulated by the ingenious labors of decades through Molien, Frobenius, Dickson, Wedderburn, and others. The notion of the ideal in several new versions again plays the decisive part. Besides it, the idea of *automorphism* proves to be rather useful, i.e., of those mappings one can perform within an algebra without destroying the internal relations. Calculative tools are discarded like, for instance, a certain determinant the non-vanishing of which Dedekind had used as a criterion for semi-simplicity; this was the more desirable as this criterion fails in some domains of rationality. In intense cooperation with Hasse and with Brauer she investigated the structure of non-commutative algebras and applied the theory by means of her *verschränktes Produkt* (cross product) to the ordinary commutative number fields and their arithmetics. The most important papers of this epoch are "Hyperkomplexe Grössen und Darstellungstheorie" ["Hypercomplex Magnitudes and Representation Theory"], 1929; "Nicht-kommutative Algebra" ["Non-Commutative Algebra"], 1933; and three smaller papers about norm rests and the principal genus theorem. Her theory of cross products was published by Hasse in connection with his investigations about the theory of cyclic algebras. A common paper by Brauer, Hasse, and Emmy Noether proving the fact that every simple algebra over an ordinary algebraic number field is cyclic in Dickson's sense, will remain a high mark in the history of algebra.

I must forego giving a picture of the content of these profound investigations. Instead, I had better try to close with a short general estimate of Emmy Noether as a mathematician and as a personality.

Her strength lay in her ability to operate abstractly with concepts. It was not necessary for her to allow herself to be led to new results on the leading strings of known concrete examples. This had the disadvantage, however, that she was sometimes but incompletely cognizant of the specific details of the more interesting applications of her general theories. She possessed a most vivid imagination, with the aid of which she could visualize remote connections; she constantly strove toward unification. In this she sought out the essentials in the known facts, brought them into order by means of appropriate general concepts, espied the vantage point from which the whole could best be surveyed, cleansed the object under consideration of superfluous dross, and thereby won through to so simple and distinct a form that the venture into new territory could be undertaken with the greatest prospect of success. This clarifying power she proved, for example, in her theory of the cross product, in which almost all the facts had already been found by Dickson and by Brauer. She possessed a strong drive toward axiomatic purity. All should be accomplished within the frame and with the aid of the intrinsic properties of the structure under investigation; nothing should be brought from without, and only invariant processes should be applied. Thus it seemed to her that the use of matrices which commute with all the elements of a given matric algebra, so often to be found in the work of Schur, was inappropriate; accordingly she used the automorphisms instead. This can be carried too far, however, as when she disdained to employ a primitive element in the development of the Galois theory. She once said:

> If one proves the equality of two numbers a and b by showing first that $a \leqq b$ and then $a \geqq b$, it is unfair; one should instead show that they are really equal by disclosing the inner ground for their equality.

Of her predecessors in algebra and number theory, Dedekind was most closely related to her. For him she felt a deep veneration. She expected her students to read Dedekind's appendices to Dirichlet's *Zahlentheorie* [*Theory of Numbers*] not only in one, but in all editions. She took a most active part in the editing of Dedekind's works; here the attempt was made to indicate, after each of Dedekind's papers, the modern development built upon his investigations.[15] Her affinity with Dedekind, who was perhaps the most typical Lower Saxon among German mathematicians, proves by a glaring example how illusory it is to associate in a schematic way race with the style of mathematical thought. In addition to Dedekind's work, that of Steinitz on the theory of abstract fields was naturally of great importance for her own work. She lived through a great flowering of algebra in Germany, toward which she contributed much. Her methods need not, however, be considered the only means of salvation. In addition to Artin and Hasse, who in some respects are akin to her, there are algebraists of a still more different stamp, such as I. Schur in Germany, Dickson and Wedderburn in America, whose achievements are certainly not behind hers in depth and significance. Perhaps her followers, in pardonable enthusiasm, have not always fully recognized this fact.

Emmy Noether was a zealous collaborator in the editing of the *Mathematische Annalen*. That this work was never explicitly recognized may have caused her some pain.

It was only too easy for those who met her for the first time, or had no feeling for her creative power, to consider her queer and to make fun at her expense. She was heavy of build and loud of voice, and it was often not easy for one to get the floor in competition with her. She preached mightily, and not as the scribes. She was a rough and simple soul, but her heart was in the right place. Her frankness was never offensive in the least degree. In everyday life she was most unassuming and utterly unselfish; she had a kind and friendly nature. Nevertheless she enjoyed the recognition paid her; she could answer with a bashful smile like a young girl to whom one had whispered a compliment. No one could contend that the Graces had stood by her cradle; but if we in Göttingen often chaffingly referred to her as "*der Noether*" (with the masculine article), it was also done with a respectful recognition of her power as a creative thinker who seemed to have broken through the barrier of sex. She possessed a rare humor and a sense of sociability; a tea in her apartments could be most pleasurable. But she was a one-sided being who was thrown out of balance by the overweight of her mathematical talent. Essential aspects of human life remained undeveloped in her, among them, I suppose, the erotic, which, if we are to believe the poets, is for many of us the strongest source of emotions, raptures, desires, and sorrows, and conflicts. Thus she sometimes gave the impression of an unwieldly child, but she was a kind-hearted and courageous being, ready to help, and capable of the deepest loyalty and affection. And of all I have known, she was certainly one of the happiest.

Comparison with the other woman mathematician of world renown, Sonya Kovalevskaya, suggests itself. Sonya had certainly the more complete personality, but was also of a much less happy nature. In order to pursue her studies Sonya had to defy the opposition of her parents, and entered into a marriage in name only, although it did not quite remain so. Emmy Noether had, as I have already indicated, neither a rebellious nature nor Bohemian leanings. Sonya possessed feminine charm, instincts, and vanity; social successes were by no means immaterial to her. She was a creature of tension and whims; mathematics made her unhappy, whereas Emmy found the greatest pleasure in her work. Sonya followed literary pursuits outside of mathematics. In her later years in Paris, as she worked feverishly on a paper to be submitted for a mathematical prize, Sonya, alluding in a letter to a certain M. with whom she was in love, wrote "The fat M. occupies all the room on my couch and in my thoughts." Such was Sonya: you see the tension between her creative mind and life with its passion and the self-mocking spirit ironically viewing her own desperate conflict. How far from Emmy's possibilities! But Emmy Noether without doubt possessed by far the greater power, the greater scientific talent.

Indeed, two traits determined above all her nature: First, the native productive power of her mathematical genius. She was not clay, pressed by the artistic hands of God into a harmonious form, but rather a chunk of human primary rock into which he had blown his creative breath of life. Second, her heart knew no malice; she did not believe in evil — indeed it never entered her mind that it could play a role among men. This was never more forcefully apparent to me than in the last stormy summer, that of 1933, which we spent together in Göttingen. The memory of her work in science and of her personality among her fellows will not soon pass away. She was a great mathematician, the greatest, I firmly believe, that her sex has ever produced, and a great woman.

Notes

[Memorial Address delivered in Goodheart Hall, Bryn Mawr College, 26 April 1935; originally published in Weyl 1935, *WGA* 3:425–444, also included in Dick 1981, 112–153 and abstracted in Catt 1935. For the text of Weyl's remarks at Noether's funeral, see Roquette 2008, 319–321.]

[1] [For Dirichlet's principle, see below, 180.]

[2] [A "Bursche" is a member of a German student fraternity or association (*Burschenschaft*), many of which were involved in efforts for German unity during the revolution of 1848.]

[3] [*Ex ungue leonem*: "From its claw [we may judge of] the lion;" for instance, Johann Bernoulli used this Latin phrase to describe the brilliance of Newton's solution to the probem of the brachistochrone, to find the shortest path in time between two given points.]

[4] [See Weyl 1950, 369–377.]

[5] [For his inauguration to his professorship in Erlangen (1872), Klein wrote a pamplet, now known as the "Erlangen Program" (Weyl sometimes uses its German spelling, "Erlanger Programm"), championing a new approach to geometry, based on his work with Sophus Lie. Klein advocated considering every geometry as a space on which a particular group of transformations acts to preserve certain invariants. For the text of Klein's program, see Klein 1921, 1:460–497 and Rowe 1985; for commentary, see Hawkins 1984, Hawkins 2000, 34–42, Yaglom 1988, 111-124,Rowe 1992, Laugwitz 1999, 246–252, Gray 2005, Gray 2008, 114–117.]

[6] [Klein 1956.]

[7] [See Brewer and Smith 1981, 152–153, and Roquette 2005.]

[8] [For an excellent account of this work, and of Noether's subsequent development of abstract algebra, see McClarty 2006 and Kleiner 2007, 91–102. For a general survey of her work, see Brewer and Dick 1981, 115–163.]

[9] [This is the celebrated "Noether principle" (called by physicists "Noether's theorem"), prepared when she came to collaborate with Hilbert in Göttingen in 1915 (in his independent attempt to reach the equations of general relativity), translated in Tavel 1971; see Byers 1996, Brading 2002, Brading and Brown 2003, Martin 2003, and Kosmann-Schwarzbach 2004.]

[10] [This rank might be translated as "unofficial associate professor" or perhaps "adjunct professor"; the eminent physicist Lise Meitner was also relegated to such a post in the 1930s. A *Lehrauftrag* is a lectureship.]

[11] [For Alexandrov's memoir, see Dick 1981, 153–179; she spent the winter of 1928–1929 lecturing at Moscow University. At the time of her death, her brother Fritz was a professor of mathematics at Tomsk University.]

[12] [A person who has completed the postdoctoral higher qualification of "habilitation" is then granted the *venia legendi* ("permission to lecture"), which allows them to be a *Privatdozent* or any other higher academic rank. See also 12, above.]

[13] [This quotation comes from the end of his essay on "Topology and Abstract Algebra"; see above, 47.]

[14] [*Teilerkettensatz*, the "ascending chain condition," is the defining property of what now are called "Noetherian rings." See Brewer and Smith 1981, 20–21.]

[15] [Van der Waerden 1975 remarks that Noether's motto was "Es steht alles schon bei Dedekind": "You can find that already in Dedekind." As one of Noether's most brilliant students, van der Waerden 1985 discusses her in detail.]

The Mathematical Way of Thinking
(1940)

By the mathematical way of thinking I mean first that form of reasoning through
which mathematics penetrates into the sciences of the external world — physics,
chemistry, biology, economics, etc., and even into our everyday thoughts about
human affairs, and secondly that form of reasoning which the mathematician,
left to himself, applies in his own field. By the mental process of thinking we try
to ascertain truth; it is our mind's effort to bring about its own enlightenment
by evidence. Hence, just as truth itself and the experience of evidence, it is
something fairly uniform and universal in character. Appealing to the light in
our innermost self, it is neither reducible to a set of mechanically applicable
rules, nor is it divided into watertight compartments like historic, philosophical,
mathematical thinking, etc. We mathematicians are no Ku Klux Klan with a
secret ritual of thinking. True, nearer the surface there are certain techniques
and differences; for instance, the procedures of fact-finding in a courtroom and
in a physical laboratory are conspicuously different. However, you should not
expect me to describe the mathematical way of thinking much more clearly than
one can describe, say, the democratic way of life.

A movement for the reform of the teaching of mathematics, which some
decades ago made quite a stir in Germany under the leadership of the great mathematician Felix Klein, adopted the slogan "functional thinking." The important
thing which the average educated man should have learned in his mathematics
classes, so the reformers claimed, is thinking in terms of *variables and functions*.
A function describes how one variable y depends on another x; or more generally, it maps one variety, the range of a variable element x, upon another (or
the same) variety. This idea of function or mapping is certainly one of the most
fundamental concepts, which accompanies mathematics at every step in theory
and application.

Our federal income tax law defines the tax y to be paid in terms of the income
a; it does so in a clumsy enough way by pasting several linear functions together,
each valid in another interval or bracket of income. An archeologist who, five
thousand years from now, shall unearth some of our income tax returns together
with relics of engineering works and mathematical books, will probably date
them a couple of centuries earlier, certainly before Galileo and Vieta.[1] Vieta was

instrumental in introducing a consistent algebraic symbolism; Galileo discovered the quadratic law of falling bodies, according to which the drop s of a body falling in a vacuum is a quadratic function of the time t elapsed since its release:

$$(1) \qquad\qquad s = \frac{1}{2}gt^2,$$

g being a constant which has the same value for each body at a given place. By this formula Galileo converted a natural law inherent in the actual motion of bodies into an a priori constructed mathematical function, and that is what physics endeavors to accomplish for every phenomenon. The law is of much better design than our tax laws. It has been designed by Nature, who seems to lay her plans with a fine sense for mathematical simplicity and harmony. But then Nature is not, as our income and excess profits tax laws are, hemmed in by having to be comprehensible to our legislators and chambers of commerce.

Right from the beginning we encounter these characteristic features of the mathematical process: 1) variables, like t and s in the formula (1), whose possible values belong to a range, here the range of real numbers, which we can completely survey because it springs from our own free construction, 2) representation of these variables by symbols, and 3) functions or a priori constructed mappings of the range of one variable t upon the range of another s. *Time* is the independent variable *kat' exochen* [par excellence].

In studying a function one should let the independent variable run over its full range. A conjecture about the mutual interdependence of quantities in nature, even before it is checked by experience, may be probed in thought by examining whether it carries through over the whole range of the independent variables. Sometimes certain simple limiting cases at once reveal that the conjecture is untenable. Leibnitz taught us by his *principle of continuity* to consider rest not as contradictorily opposed to motion, but as a limiting case of motion. Arguing by continuity he was able a priori to refute the laws of impact proposed by Descartes. Ernst Mach gives this prescription: "After having reached an opinion for a special case, one gradually modifies the circumstances of this case as far as possible, and in so doing tries to stick to the original opinion as closely as one can. There is no procedure which leads more safely and with greater mental economy to the simplest interpretation of all natural events." Most of the variables with which we deal in the analysis of nature are continuous variables like time, but although the word seems to suggest it, the mathematical concept is not restricted to this case. The most important example of a discrete variable is given by the sequence of natural numbers or integers 1, 2, 3,... Thus the number of divisors of an arbitrary integer n is a function of n.

In Aristotle's logic one passes from the individual to the general by exhibiting certain abstract features in a given object and discarding the remainder, so that two objects fall under the same concept or belong to the same genus if they have those features in common. This descriptive classification, e.g., the description of plants and animals in botany and zoology, is concerned with the actual existing objects. One might say that Aristotle thinks in terms of substance and accident, while the functional idea reigns over the formation of mathematical concepts.[2] Take the notion of ellipse. Any ellipse in the x-y-plane is a set E of points (x, y) defined by a quadratic equation

$$ax^2 + 2bxy + cy^2 = 1$$

whose coefficients a, b, c satisfy the conditions

$$a > 0, \quad c > 0, \quad ac - b^2 > 0.$$

The set E depends on the coefficients a, b, c; we have a function $E(a, b, c)$ which gives rise to an individual ellipse by assigning definite values to the variable coefficients a, b, c. In passing from the individual ellipse to the general notion one does not discard any specific difference, one rather makes certain characteristics (here represented by the coefficients) variable over an a priori surveyable range (here described by the inequalities). The notion thus extends over all *possible*, rather than over all *actually existing*, specifications.[3]

From these preliminary remarks about functional thinking I now turn to a more systematic argument. Mathematics is notorious for the thin air of abstraction in which it moves. This bad reputation is only half deserved. Indeed, the first difficulty the man in the street encounters when he is taught to think mathematically is that he must learn to look things much more squarely in the face; his belief in words must be shattered; he must learn to think more concretely. Only then will he be able to carry out the second step, the step of abstraction where intuitive ideas are replaced by purely symbolic construction.

About a month ago I hiked around Longs Peak in the Rocky Mountain National Park with a boy of twelve, Pete. Looking up at Longs Peak he told me that they had corrected its elevation and that it is now 14,255 feet instead of 14,254 feet last year I stopped a moment asking myself what this could mean to the boy, and should I try to enlighten him by some Socratic questioning. But I spared Pete the torture, and the comment then withheld, will now be served to you. Elevation is elevation above sea level. But there is no sea under Longs Peak. Well, in idea one continues the actual sea level under the solid continents. But how does one construct this ideal closed surface, the geoid, which coincides with the surface of the oceans over part of the globe? If the surface of the ocean were strictly spherical, the answer would be clear. However, nothing of this sort is the case. At this point dynamics comes to our rescue. Dynamically the sea level is a surface of constant potential $\phi = \phi_0$; more exactly ϕ denotes the gravitational potential of the earth, and hence the difference of ϕ at two points P, P' is the work one must put into a small body of mass 1 to transfer it from P to P'. Thus it is most reasonable to define the geoid by the dynamical equation $\phi = \phi_0$. If this constant value of ϕ fixes the elevation zero, it is only natural to define any fixed altitude by a corresponding constant value of ϕ, so that a peak P is called higher than P' if one gains energy by flying from P to P'. The geometric concept of altitude is replaced by the dynamic concept of potential or energy. Even for Pete, the mountain climber, this aspect is perhaps the most important: the higher the peak the greater — *ceteris paribus* – the mechanical effort in climbing it By closer scrutiny one finds that in almost every respect the potential is the relevant factor. For instance the barometric measurement of altitude is based on the fact that in an atmosphere of given constant temperature the potential is proportional to the logarithm of the atmospheric pressure, whatever the nature of the gravitational field. Thus atmospheric pressure, generally speaking, indicates potential and not altitude. Nobody who has learned that the earth is round and the vertical direction is not an intrinsic geometric property of space but the direction of gravity should be surprised that he is forced to discard the geometric idea of altitude in favor of the dynamically more concrete idea of potential. Of course there is a relationship to geometry: In a region of space so small that one can consider the force of gravity as constant throughout this region, we have a fixed vertical direction, and potential differences are proportional to differences of altitude measured in that direction. Altitude, height, is a word which has a clear meaning when I ask how high the ceiling of this room is above its floor. The meaning gradually loses precision when we apply it to the relative altitudes of mountains in a wider and wider region. It dangles in the air when we extend it to the whole globe, unless we support it by the dynamical concept of potential. Potential is more concrete than altitude because it is generated by and dependent on the mass distribution of the earth.

Words are dangerous tools. Created for our everyday life they may have their good meanings under familiar limited circumstances, but Pete and the man in the street are inclined to extend them to wider spheres without bothering about whether they then still have a sure foothold in reality. We are witnesses of the disastrous effects of this witchcraft of words in the political sphere where all words have a much vaguer meaning and human passion so often drowns the voice of reason. The scientist must thrust through the fog of abstract words to reach the concrete rock of reality. It seems to me that the science of economics has a particularly hard job, and will still have to spend much effort, to live up to this principle. It is, or should be, common to all sciences, but physicists and mathematicians have been forced to apply it to the most fundamental concepts where the dogmatic resistance is strongest, and thus it has become their second nature. For instance, the first step in explaining relativity theory must always consist in shattering the dogmatic belief in the temporal terms past, present, future. You cannot apply mathematics as long as words still becloud reality.

I return to relativity as an illustration of this first important step preparatory to mathematical analysis, the step guided by the maxim, "Think concretely." As the root of the words *past, present, future*, referring to time, we find something much more tangible than time, namely, the causal structure of the universe. Events are localized in space and time; an event of small extension takes place at a space-time or world point, a here-now. After restricting ourselves to events on a plane E we can depict the events by a graphic timetable in a three-dimensional diagram with a horizontal E plane and a vertical t axis on which time t is plotted. A world point is represented by a point in this picture, the motion of a small body by a world line, the propagation of light with its velocity c radiating from a light signal at the world point O by a vertical straight circular cone with vertex at O (light cone). The *active future* of a given world point O, here-now, contains all those events which can still be influenced by what happens at O, while its *passive past* consists of all those world points from which any influence, any message, can reach O. I here-now can no longer change anything that lies outside the active future; all events of which I here-now can have knowledge by direct observation or any records thereof necessarily lie in the passive past. We interpret the words past and future in this causal sense where they express something very real and important, the causal structure of the world.

The new discovery at the basis of the theory of relativity is the fact that no effect may travel faster than light. Hence while we formerly believed that active future and passive past bordered on each other along the cross-section of present, the horizontal plane $t = $ const. going through O, Einstein taught us that the active future is bounded by the forward light cone and the passive past by its backward continuation.

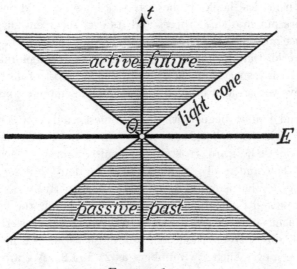

FIGURE 1

Active future and passive past are separated by the part of the world lying between these cones, and with this part I am here-now not at all causally connected. The essential positive content of relativity theory is this new insight into the causal structure of the universe. By discussing the various interpretations of such a simple question as whether two men, say Bill on earth and Bob on Sirius, are contemporaries, as to whether it means that Bill can send a message to Bob, or Bob a message to Bill, or even that Bill can communicate with Bob by sending a message and receiving an answer, etc., I often succeed soon in accustoming my listener to thinking in terms of causal rather than his wonted temporal structure. But when I tell him that the causal structure is not a stratification by horizontal layers $t = $ const., but that active future and passive past are of cone-like shape with an interstice between, then some will discern dimly what I am driving at, but every honest listener will say: Now you draw a figure, you speak in pictures; how far does the simile go, and what is the naked truth to be conveyed by it? Our popular writers and news reporters, when they have to deal with physics, indulge in similes of all sorts; the trouble is that they leave the reader helpless in finding out how far these pungent analogies cover the real issue, and therefore more often lead him astray than enlighten him. In our case one has to admit that our diagram is no more than a picture, from which, however, the real thing emerges as soon as we replace the intuitive space in which our diagrams are drawn by its construction in terms of sheer symbols. Then the phrase that the world is a four-dimensional continuum changes from a figurative form of speech into a statement of what is literally true. At this second step the mathematician turns abstract, and here is the point where the layman's understanding most frequently breaks off: the intuitive picture must be exchanged for a symbolic construction. "By its geometric and later by its purely symbolic construction," says Andreas Speiser, "mathematics shook off the fetters of language, and one who knows the enormous work put into this process and its ever recurrent surprising successes cannot help feeling that mathematics today is more efficient in its sphere of the intellectual world, than the modern languages in their deplorable state or even music are on their respective fronts." I shall spend most of my time today in an attempt to give you an idea of what this magic of symbolic construction is.[4]

To that end I must begin with the simplest, and in a certain sense most profound, example: the natural numbers or *integers* by which we *count* objects. The symbols we use here are strokes put one after another. The objects may disperse, "melt, thaw and resolve themselves into a dew," but we keep this record of their number.[5] What is more, we can by a constructive process decide for two numbers represented through such symbols which one is the larger, namely by checking one against the other, stroke by stroke. This process reveals differences not manifest in direct observation, which in most instances is incapable of distinguishing between even such low numbers as 21 and 22. We are so familiar with these miracles which the number symbols perform that we no longer wonder at them. But this is only the prelude to the mathematical step proper. We do not leave it to chance which numbers we shall meet by counting actual objects, but we generate the open sequence of *all possible* numbers which starts with 1 (or 0) and proceeds by adding to any number symbol n already reached one more stroke, whereby it changes into the following number n'. As I have often said before, being is thus projected onto the background of the possible, or more precisely onto a manifold of possibilities which unfolds by iteration and is open into infinity. Whatever number n we are given, we always deem it possible to pass to the next n'. "Number goes on." This intuition of the "ever one more," of the open countable infinity, is basic for all mathematics.[6] It gives birth to the simplest example of what I termed above an a priori surveyable range of variability. According to this process by which the integers are created, functions of an argument ranging over all integers n are to be defined by so-called complete induction, and statements holding for all n are to be proved in the same fashion. The principle of this inference by complete induction is as follows.[7] In order to show that every number n has a certain property V it is sufficient to make sure of two things:

1. 0 has this property;
2. If n is any number which has the property V, then the next number n' has the property V.

It is practically impossible, and would be useless, to write out in strokes the symbol of the number 10^{12}, which the Europeans call a billion and we in this country, a thousand billions. Nevertheless we talk about spending more than 10^{12} cents for our defense program, and the astronomers are still ahead of the financiers. In July the *New Yorker* carried this cartoon: man and wife reading the newspaper over their breakfast and she looking up in puzzled despair: "Andrew, how much is seven hundred billion dollars?" A profound and serious question, lady! I wish to point out that only by passing through the *infinite* can we attribute any significance to such figures. 12 is an abbreviation of

$$10^{12} = \frac{|\quad|\quad|\quad|\quad|\quad|\quad|\quad|\quad|\quad|\quad|\quad|}{10 \cdot 10 \cdot 10 \cdot 10 \cdot 10 \cdot 10 \cdot 10 \cdot 10 \cdot 10 \cdot 10 \cdot 10 \cdot 10}$$

can not be understood without defining the function $10 \cdot n$ for all n and this is done through the following definition by complete induction:

$$10 \cdot 0 = 0,$$

$$10 \cdot n' = (10 \cdot n)^{''''''''''}$$

The dashes constitute the explicit symbol for 10, and, as previously, each dash indicates transition to the next number. Indian, in particular Buddhist, literature indulges in the possibilities of fixing stupendous numbers by the decimal system of numeration which the Indians invented, i.e., by a combination of sums, products and powers. I mention also Archimedes's treatise "On the counting of sand," and Professor Kasner's Googolplex in his recent popular book on *Mathematics and the Imagination*.[8]

Our conception of *space* is, in a fashion similar to that of natural numbers, depending on a constructive grip on all *possible* places. Let us consider a metallic disk in a plane E. Places on the disk can be marked *in concreto* by scratching little crosses on the plate. But relatively to two axes of coordinates and a standard length scratched into the plate we can also put ideal marks in the plane outside the disk by giving the numerical values of their two coordinates. Each coordinate varies over the a priori constructed range of real numbers. In this way astronomy uses our solid earth as a base for plumbing the sidereal spaces. What a marvelous feat of imagination when the Greeks first constructed the shadows which earth and moon, illumined by the sun, cast in empty space and thus explained the eclipses of sun and moon! In analyzing a continuum, like space, we shall here proceed in a somewhat more general manner than by measurement of coordinates and adopt the *topological* viewpoint, so that two continua arising one from the other by continuous deformation are the same to us. Thus the following exposition is at the same time a brief introduction to an important branch of mathematics, topology.

The symbols for the localization of points on the one-dimensional continuum of a straight line are the *real numbers*. I prefer to consider a closed one-dimensional continuum, the circle. The most fundamental statement about a continuum is that it may be divided into parts. We catch all the points of a continuum by spanning a net of division over it, which we refine by repetition of a definite process of subdivision ad infinitum. Let S be any division of the circle into a number of arcs, say l arcs. From S we derive a new division S' by the process of *normal subdivision*, which consists in breaking each arc into two. The number of arcs in S' will then be $2l$. Running around the circle in a definite sense (orientation) we may distinguish the two pieces, in the order in which we meet them, by the marks 0 and 1; more explicitly, if the arc is denoted by a symbol α then these two pieces are designated as α_0 and α_1. We start with the division S_0 of the circle into two arcs $+$ and $-$; either is topologically a cell, i.e., equivalent to a segment. We then iterate the process of normal subdivision and thus obtain $S_0', S_0'', \ldots$, seeing to it that the refinement of the division ultimately pulverizes the whole circle. If we had not renounced the use of metric properties we could decree that the normal sub-division takes place by cutting each arc into two *equal* halves. We introduce no such fixation; hence the actual performance of the process involves a wide measure of arbitrariness. However, the *combinatorial scheme* according to which the parts reached at any step border on each other, and according to which the division progresses, is unique and perfectly fixed. Mathematics cares for this symbolic scheme only. By our notation the parts occurring at the consecutive divisions are *catalogued* by symbols of this type

$$+.011010001$$

with $+$ or $-$ before the dot and all following places occupied by either 0 or 1. We see that we arrive at the familiar symbols of binary (not decimal) fractions. A point is caught by an infinite sequence of arcs of the consecutive divisions such that each arc arises from the preceding one by choosing one of the two pieces into which it breaks by the next normal subdivision, and the point is thus fixed by an

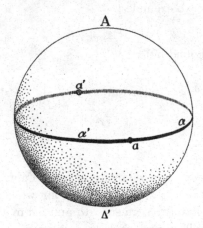

FIGURE 2

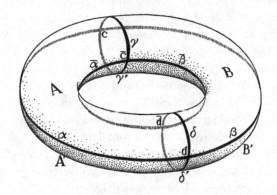

FIGURE 3

infinite binary fraction. Let us try to do something similar for two-dimensional continua, e.g., for the surface of a sphere or a torus. The figures [2, 3] show how we may cast a very coarse net over either of them, the one consisting of two, the other of four meshes; the globe is divided into its upper and lower halves by the equator, the torus is welded together from four rectangular plates. The meshes are two-dimensional cells, or briefly, 2-cells which are topologically equivalent to a circular disk. The combinatorial description is facilitated by introducing also the vertices and edges of the division, which are 0- and 1-cells. We attach arbitrary symbols to them and state in symbols for each 2-cell which 1-cells bound it, and for each 1-cell by which 0-cells it is bounded. We then arrive at a *topological scheme* S. Here are our two examples:

Sphere. $A \to \alpha, \alpha'$. $A' \to \alpha, \alpha'$. $\alpha \to a, a'$. $\alpha' \to a, a'$.
          ($\to$ means: bound by)

Torus. $A \to \alpha, \bar{\alpha}, \gamma, \delta$. $A' \to \alpha, \bar{\alpha}, \gamma', \delta'$.
          $B \to \beta, \bar{\beta}, \gamma, \delta$. $B' \to \beta, \bar{\beta}, \gamma', \delta'$.
          $\alpha \to c, d$. $\bar{\alpha} \to \bar{c}, \bar{d}$. $\beta \to c, d$. $\bar{\beta} \to \bar{c}, \bar{d}$.
          $\gamma \to c, \bar{c}$. $\gamma' \to c, \bar{c}$. $\delta \to d, \bar{d}$. $\delta' \to d, \bar{d}$.

From this initial stage we proceed by iteration of a universal process of normal subdivision: On each 1-cell $\alpha = ab$ we choose a point which serves as a new vertex α and divides the 1-cell into two segments αa and αb; in each 2-cell A we choose a point A and cut the cell into triangles by joining the newly created vertex A with the old and new vertices on its bounding 1-cells by lines within the 2-cell. Just as in elementary geometry we denote the triangles and their sides by means of their vertices.

The figure [4] shows a pentagon before and after subdivision; the triangle $A\beta c$ is bounded by the 1-cells $\beta c, A\beta, Ac$, the 1-cell Ac for instance by the vertices c and A. We arrive at the following general purely symbolic description of the process by which the subdivided scheme S' is derived from a given topological scheme S. Any symbol $e_2 e_1 e_0$ made up by the symbols of a 2-cell e_2, a 1-cell e_1 and a 0-cell e_0 in S such that e_2 is bounded by e_1 and e_1 bounded by e_0 represents a 2-cell e_2' of S'. This 2-cell $e_2' = e_2 e_1 e_0$ in S' is part of the 2-cell e_2 in S. The symbols of cells in S' which bound a given cell are derived from its symbol by dropping any one of its constituent letters. Through iteration of this symbolic process the initial scheme S_0 gives rise to a sequence of derived schemes $S_0', S_0'', S_0''', \cdots$. What we have done is nothing else than devise a systematic cataloguing of the parts created by consecutive subdivisions. A *point* of our continuum is caught by a sequence

$$(2) \qquad\qquad\qquad e\, e'\, e'' \cdots$$

which starts with a 2-cell e of S_0 and in which the 2-cell $e^{(n)}$ of the scheme $S^{(n)}$ is followed by one of the 2-cells $e^{(n+1)}$ of $S^{(n+1)}$ into which $e^{(n)}$ breaks up by our subdivision. (To do full justice to the inseparability of parts in a continuum this description ought to be slightly altered. But for the present purposes our simplified description will do.) We are convinced that not only may each point be caught by such a sequence (Eudoxus), but that an arbitrarily constructed sequence of this sort always catches a point (Dedekind, Cantor). The fundamental concepts of *limit, convergence,* and *continuity* follow in the wake of this construction.

We now come to the decisive step of mathematical abstraction: we forget about what the symbols stand for. The mathematician is concerned with the

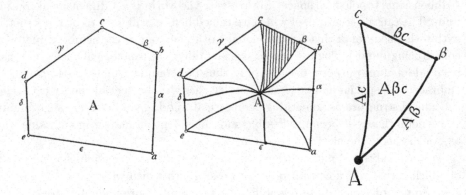

FIGURE 4

catalogue alone; he is like the man in the catalogue room who does not care what books or pieces of an intuitively given manifold the symbols of his catalogue denote. He need not be idle; there are many operations which he may carry out with these symbols, without ever having to look at the things they stand for. Thus, replacing the points by their symbols (2) he turns the given manifold into a *symbolic construct* which we shall call the *topological space* $\{S_0\}$ because it is based on the scheme S_0 alone.

The details are not important; what matters is that once the initial finite symbolic scheme S_0 is given we are carried along by an absolutely rigid symbolic construction which leads from S_0 to S_0', from S_0' to S_0'', etc. The idea of iteration, first encountered with the natural numbers, again plays a decisive role. The realization of the symbolic scheme for a given manifold, say a sphere or a torus, as a scheme of consecutive divisions involves a wide margin of arbitrariness restricted only by the requirement that the pattern of the meshes ultimately becomes infinitely fine everywhere. About this point and the closely affiliated requirement that each 2-cell has the topological structure of a circular disk, I must remain a bit vague. However, the mathematician is not concerned with applying the scheme or catalogue to a given manifold, but only with the scheme itself, which contains no haziness whatsoever. And we shall presently see that even the physicist need not care greatly about that application. It was merely for heuristic purposes that we had to go the way from manifold through division to pure symbolism. In the same purely symbolic way we can evidently construct not only 1- and 2- but also $3, 4, 5, \ldots$-dimensional manifolds. An n-dimensional scheme S_0 consists of symbols distinguished as 0, 1, 2, $\ldots$, n-cells and associates with each i-cell e_i $(i = 1, 2, \ldots, n)$ certain $(i - 1)$-cells of which one says that they bound e_i. It is clear how the process of normal subdivision carries over. *A certain such 4-dimensional scheme can be used for the localization of events,* of all possible here-nows; physical quantities which vary in space and time are functions of a variable point ranging over the corresponding symbolically constructed

4-dimensional topological space. In this sense the world *is* a 4-dimensional continuum. The causal structure, of which we talked before, will have to be constructed within the medium of this 4-dimensional world, i.e., out of the symbolic material constituting our topological space. Incidentally the topological viewpoint has been adopted on purpose, because only thus our frame becomes wide enough to embrace both special and general relativity theory. The special theory envisages the causal structure as something geometrical, rigid, given once for all, while in the general theory it becomes flexible and dependent on matter in the same way as, for instance, the electromagnetic field.

In our analysis of nature we reduce the phenomena to simple elements each of which varies over a certain range of possibilities which we can survey a priori because we construct these possibilities a priori in a purely combinatorial fashion from some purely symbolic material. The manifold of space-time points is one, perhaps the most basic one, of these constructive elements of nature. We dissolve light into plane polarized monochromatic light beams with few variable characteristics like wavelength which varies over the symbolically constructed continuum of real numbers. Because of this a priori construction we speak of a *quantitative* analysis of nature; I believe the word quantitative, if one can give it a meaning at all, ought to be interpreted in this wide sense. The power of science, as witnessed by the development of modern technology, rests upon the combination of a priori symbolic construction with systematic experience in the form of planned and reproducible reactions and their measurements. As material for the a priori construction, Galileo and Newton used certain features of reality like space and time which they considered as objective, in opposition to the subjective sense qualities, which they discarded. Hence the important role which geometric figures played in their physics. You probably know Galileo's words in the *Saggiatore* [*Assayer*] where he says that no one can read the great book of nature "unless he has mastered the code in which it is composed, that is, the mathematical figures and the necessary relations between them."[9] Later we have learned that none of these features of our immediate observation, not even space and time, have a right to survive in a pretended truly objective world, and thus have gradually and ultimately come to adopt a purely symbolic combinatorial construction.

While a set of objects determines its number unambiguously, we have observed that a scheme of division S_0 with its consecutive derivatives $S_0', S_0'', \cdots$ can be established on a given manifold in many ways involving a wide margin of arbitrariness. But the question whether two schemes,

$$S_0, S_0', S_0'', \cdots \text{ and } T_0, T_0', T_0'' \cdots$$

are fit to describe the same manifold is decidable in a purely mathematical way: it is necessary and sufficient that the two topological spaces $\{S_0\}$ and $\{T_0\}$ can be mapped one upon the other by a continuous one-to-one transformation – a condition which ultimately boils down to a certain relationship called isomorphism between the two schemes S_0 and T_0. (Incidentally the problem of establishing the criterion of isomorphism for two finite schemes in finite combinatorial form is one of the outstanding unsolved mathematical problems.) The connection between a given continuum and its symbolic scheme inevitably carries with it this notion of *isomorphism*; without it and without our understanding that isomorphic schemes are to be considered as not intrinsically different, no more than congruent figures in geometry, the mathematical concept of a topological space would be incomplete. Moreover it will be necessary to formulate precisely the conditions which every topological scheme is required to satisfy. For instance, one such condition demands that each 1-cell be bounded by exactly *two* 0-cells.

I can now say a little more clearly why the physicist is almost as disinterested as the mathematician in the particular way how a certain combinatorial scheme of consecutive divisions is applied to the continuum of here-nows which we called the world. Of course, somehow our theoretical constructions must be put in contact with the observable facts. The historic development of our theories proceeds by heuristic arguments over a long and devious road and in many steps from experience to construction. But systematic exposition should go the other way: first develop the theoretical scheme without attempting to define individually by appropriate measurements the symbols occurring in it as space-time coordinates, electromagnetic field strengths, etc., then describe, as it were in one breath, the contact of the whole system with observable facts. The simplest example I can find is the observed angle between two stars.[10] The symbolic construct in the medium of the 4-dimensional world from which theory determines and predicts the value of this angle includes: (1) the world-lines of the two stars, (2) the causal structure of the universe, (3) the world position of the observer and the direction of his world line at the moment of observation. But a continuous deformation, a one-to-one continuous transformation of this whole picture, does not affect the value of the angle. *Isomorphic pictures lead to the same results concerning observable facts.* This is, in its most general form, the *principle of relativity*. The arbitrariness involved in our ascent from the given manifold to the construct is expressed by this principle for the opposite descending procedure, which the systematic exposition should follow.

So far we have endeavored to describe how a mathematical construct is distilled from the given raw material of reality. Let us now look upon these products of distillation with the eye of a pure mathematician. One of them is the sequence of natural numbers and the other the general notion of a topological space $\{S_0\}$ into which a topological scheme S_0 develops by consecutive derivations $S_0, S_0', S_0'', \cdots$. In both cases *iteration* is the most decisive feature. Hence all our reasoning must be based on evidence concerning that completely transparent process which generates the natural numbers, rather than on any principles of formal logic like syllogism, etc. The business of the constructive mathematician is *not* to draw logical conclusions. Indeed his arguments and propositions are merely an accompaniment of his actions, his carrying out of constructions. For instance, we run over the sequence of integers 0, 1, 2,... by saying alternatingly even, odd, even, odd, etc., and in view of the possibility of this inductive construction which we can extend as far as we ever wish, we formulate the general arithmetical proposition: "Every integer is even or odd." Besides the idea of iteration (or the sequence of integers) we make constant use of mappings or of the functional idea. For instance, just now we have defined a function $\pi(n)$, called parity, with n ranging over all integers and π capable of the two values 0 (even) and 1 (odd), by this induction:

$$\pi(0) = 0;$$

$$\pi(n') = 1 \text{ if } \pi(n) = 0, \qquad \pi(n') = 0 \text{ if } \pi(n) = 1.$$

Structures such as the topological schemes are to be studied in the light of the idea of *isomorphism*. For instance, when it comes to introducing operators τ which carry any topological scheme S into a topological scheme $\tau(S)$ one should pay attention only to such operators or functions τ for which isomorphism of S and R entails isomorphism for $\tau(S)$ and $\tau(R)$.

Up to now I have emphasized the constructive character of mathematics. In our actual mathematics there vies with it the non-constructive *axiomatic method*. Euclid's axioms of geometry are the classical prototype. Archimedes employs the method with great acumen and so do later Galileo and Huyghens in erecting the science of mechanics. One defines all concepts in terms of a few undefined basic concepts and deduces all propositions from a number of basic propositions, the axioms, concerning the basic concepts. In earlier times authors were inclined to claim a priori evidence for their axioms; however this is an epistemological aspect which does not interest the mathematician. Deduction takes place according to the principles of formal logic, in particular it follows the syllogistic scheme. Such a treatment *more geometrico* was for a long time considered the ideal of every science. Spinoza tried to apply it to ethics. For the mathematician the meaning of the words representing the basic concepts is irrelevant; any interpretation of them which fits, i. e., under which the axioms become true, will be good, and all the propositions of the discipline will hold for such an interpretation because they are all logical consequences of the axioms. Thus n-dimensional Euclidean geometry permits another interpretation where points are distributions of electric current in a given circuit consisting of n branches which connect at certain branch points. For instance, the problem of determining that distribution which results from given electromotoric forces inserted in the various branches of the net corresponds to the geometric construction of orthogonal projection of a point upon a linear subspace. From this standpoint mathematics treats of relations in a hypothetical-deductive manner without binding itself to any particular material interpretation. It is not concerned with the *truth* of axioms, but only with their *consistency*; indeed inconsistency would a priori preclude the possibility of our ever coming across a fitting interpretation. "Mathematics is the science which draws necessary conclusions," says B. Peirce in 1870, a definition which was in vogue for decades after. To me it seems that it renders very scanty information about the real nature of mathematics, and you are at present watching my struggle to give a fuller characterization. Past writers on the philosophy of mathematics have so persistently discussed the axiomatic method that I don't think it necessary for me to dwell on it at any greater length, although my exposition thereby becomes somewhat lopsided.

However I should like to point out that since the axiomatic attitude has ceased to be the pet subject of the methodologists its influence has spread from the roots to all branches of the mathematical tree. We have seen before that topology is to be based on a full enumeration of the axioms which a *topological scheme* has to satisfy. One of the simplest and most basic axiomatic concepts which penetrates all fields of mathematics is that of *group*. Algebra with its *"fields," "rings,"* etc., is today from bottom to top permeated by the axiomatic spirit. Our portrait of mathematics would look a lot less hazy, if time permitted me to explain these mighty words which I have just uttered, group, field and ring. I shall not try it, as little as I have stated the axioms characteristic for a topological scheme. But such notions and their kin have brought it about that modern mathematical research often is a dexterous blending of the constructive and the axiomatic procedures. Perhaps one should be content to note their mutual interlocking. But temptation is great to adopt one of these two views as the genuine primordial way of mathematical thinking, to which the other merely plays a subservient role, and it is possible indeed to carry this standpoint through consistently whether one decides in favor of construction or axiom.

Let us consider the first alternative. Mathematics then consists primarily of construction. The occurring sets of axioms merely *fix the range of variables entering into the construction*. I shall explain this statement a little further by our examples of causal structure and topology. According to the special theory of relativity the causal structure is once for all fixed and can therefore be explicitly constructed. Nay, it is reasonable to construct it together with the topological medium itself, as for instance a circle together with its metric structure is obtained by carrying out the normal subdivision by cutting each arc into two equal halves. In the general theory of relativity, however, the causal structure is something flexible; it has only to satisfy certain axioms derived from experience which allow a considerable measure of free play. But the theory goes on by establishing laws of nature which connect the flexible causal structure with other flexible physical entities, distribution of masses, electromagnetic field, etc., and these laws in which the flexible things figure as variables are in their turn *constructed* by the theory in an explicit a priori way. Relativistic cosmology asks for the topological structure of the universe as a whole, whether it is open or closed, etc. Of course the topological structure can not be flexible as the causal structure is, but one must have a free outlook on all topological possibilities before one can decide by the testimony of experience which of them is realized by our actual world. To that end one turns to topology. There the topological scheme is bound only by certain axioms; but the topologist derives numerical characters from, or establishes universal connections between, arbitrary topological schemes, and again this is done by explicit construction into which the arbitrary schemes enter as variables. Wherever axioms occur, they ultimately serve to describe the range of variables in explicitly constructed functional relations.

So much about the first alternative. We turn to the opposite view, which subordinates construction to axioms and deduction, and holds that mathematics consists of systems of axioms freely agreed upon, and their necessary conclusions. In a completely axiomatized mathematics construction can come in only secondarily as construction of examples, thus forming the bridge between pure theory and its applications. Sometimes there is only *one* example because the axioms, at least up to arbitrary isomorphisms, determine their object uniquely; then the demand for translating the axiomatic set-up into an explicit construction becomes especially imperative. Much more significant is the remark that an axiomatic system, although it refrains from constructing the mathematical *objects*, constructs the mathematical *propositions* by combined and iterated application of logical rules. Indeed, drawing conclusions from given premises proceeds by certain logical rules which since Aristotle's day one has tried to enumerate completely. Thus on the level of propositions, the axiomatic method is undiluted constructivism. David Hilbert has in our day pursued the axiomatic method to its bitter end where all mathematical propositions, including the axioms, are turned into formulas and the game of deduction proceeds from the axioms by rules which take no account of the meaning of the formulas. The mathematical game is played in silence, without words, like a game of chess. Only the rules have to be explained and communicated in words, and of course any arguing about the possibilities of the game, for instance about its consistency, goes on in the medium of words and appeals to evidence.

If carried so far, the issue between explicit construction and implicit definition by axioms ties up with the last foundations of mathematics. Evidence based on construction refuses to support the principles of Aristotelian logic when these are applied to existential and general propositions in infinite fields like the sequence of integers or a continuum of points. And if the logic of the infinite is taken into account, it seems impossible to axiomatize adequately even the most primitive process, the transition $n \to n'$ from an integer to its follower n'. As K. Gödel has shown, there will always be constructively evident arithmetical propositions which cannot be deduced from the axioms however you formulate them, while at the same time the axioms, riding roughshod over the subtleties of the constructive infinite, go far beyond what is justifiable by evidence. We are not surprised that a concrete chunk of nature, taken in its isolated phenomenal existence, challenges our analysis by its inexhaustibility and incompleteness; it is for the sake of completeness, as we have seen, that physics projects what is given onto the background of the possible. However, it is surprising that a construct created by mind itself, the sequence of integers, the simplest and most diaphanous thing for the constructive mind, assumes a similar aspect of obscurity and deficiency when viewed from the axiomatic angle. But such is the fact; which casts an uncertain light upon the relationship of evidence and mathematics. In spite, or because, of our deepened critical insight we are today less sure than at any previous time of the ultimate foundations on which mathematics rests.

My purpose in this address has not been to show how the inventive mathematical intellect works in its manifold manifestations, in calculus, geometry, algebra, physics, etc., although that would have made a much more attractive picture. Rather, I have attempted to make visible the sources from which all these manifestations spring. I know that in an hour's time I can have succeeded only to a slight degree. While in other fields brief allusions are met by ready understanding, this is unfortunately seldom the case with mathematical ideas. But I should have completely failed if you had not realized at least this much, that mathematics, in spite of its age, is not doomed to progressive sclerosis by its growing complexity, but is still intensely alive, drawing nourishment from its deep roots in mind and nature.

Notes

[An address delivered at the Bicentennial Celebration Conference of the University of Pennsylvania, 17 September 1940; the text here follows Weyl 1940; note that the text reprinted in *WGA* 3:710–718 is garbled.]

[1] [François Viète (also known in his Latinized name Franciscus Vieta), 1540–1603, was an important figure in the development of modern algebra, who introduced alphabetic symbolic notation for unknowns and coefficients as well as the general rules for manipulating them; see Klein 1968, 150–185, and Pesic 1997.]

[2] [The English words customarily used to render Aristotle's concepts are rather opaque; Weyl may here presume that the reader knows, from the meanings of the underlying Greek words, that "substance" (*ousia*) denotes the common essence (say of a biological genus), where "accident" (*sumbebekos*) denotes a quality of an individual member of that genus that does not specifically reflect its underlying essence. For example, some quality of a tree may reflect its particular "accidental" circumstances (like the local climate or water supply) rather than the essential qualities by which it can be distinguished from trees of other genera.]

[3] Compare about this contrast Ernst Cassirer, *Substanzbegriff und Funktionsbegriff*, 1910 [Cassirer 1953], and my critical remark, *Philosophie der Mathmetik und Naturwissenschaft*, 1923, p. 111. [Weyl 2009b, 149–150, 216]

[4] [Weyl gives further treatment of the philosophic aspect of symbols in his essay on "The Unity of Knowledge," in Weyl 2009a, 194–203.]

[5] [Shakespeare, *Hamlet*, I.ii.129–130.]

[6] [Given the pervasive influence of Goethe on Weyl and his contemporaries, one wonders whether he intends an echo of Faust, Part I, line 1700: "Verweile doch, du bist so schön" ("Stay a while! You are so lovely"), in which Faust insists that he will never "grow stagnant nor be a slave" (line 1710). Weyl's concept of number may thus be Faustian in character.]

[7] [For the distinction between Weyl's use of the terms "complete induction" and the contemporary meaning of this phrase, see above 31n6.]

[8] [Weyl refers to Kasner and Newman 1940/2001, whose author, in conversations with children, devised the names "google" for 10^{100} and the less well-known term "googleplex" for $10^{\text{google}} = 10^{10^{100}}$.]

[9] [Galileo 1957, 237–238.]

[10] [This example dates back to Weyl's 1927 writings; see Weyl 2009a, 30.]

David Hilbert 1862-1943
(1944)

David Hilbert, upon whom the world looked during the last decades as the greatest of the living mathematicians, died in Göttingen, Germany, on 14 February
1943. At the age of eighty-one he succumbed to a compound fracture of the thigh
brought about by a domestic accident.

Hilbert was born on 23 January 1862, in the city of Königsberg in East Prussia. He was descended from a family which had long been settled there and
had brought forth a series of physicians and judges. During his entire life he
preserved uncorrupted the Baltic accent of his home. For a long time Hilbert
remained faithfully attached to the town of his forebears, and well deserved its
honorary citizenship which was bestowed upon him in his later years. It was at the
University of Königsberg that he studied, where in 1884 he received his doctor's
degree, and where in 1886 he was admitted as *Privatdozent*; there, moreover, he
was appointed *Ausserordentlicher Professor*, in 1892, succeeding his teacher and
friend Adolf Hurwitz, and in the following year advanced to a full professorship.[1]
The continuity of this Königsberg period was interrupted only by a semester's
studies at Erlangen, and by a travelling scholarship during the year before his
habilitation, which took him to Felix Klein at Leipzig and to Paris where he was
attracted mainly to Charles Hermite. It was on Klein's initiative that Hilbert
was called to Göttingen in 1895; there he remained until the end of his life. He
retired in the year 1930.

In 1928 he was elected to foreign membership of the Royal Society.

Beginning in his student years at Königsberg a close friendship existed between him and Hermann Minkowski, his junior by two years, and it was with deep
satisfaction that in 1902 he succeeded in bringing Minkowski also to Göttingen.
Only too soon did the close collaboration of the two friends end with Minkowski's
death in 1909. Hilbert and Minkowski were the real heroes of the great and
brilliant period, unforgettable to those who lived through it, which mathematics
experienced during the first decade of this century in Göttingen. Klein ruled over
it like a distant god, *divus Felix*, from above the clouds; the peak of his mathematical productivity lay behind him. Among the authors of the great number of
valuable dissertations which in these fruitful years were written under Hilbert's
guidance we find many Anglo-Saxon names, names of men who subsequently

have played a considerable role in the development of American mathematics. The physical set-up within which this free scientific life unfolded was quite modest. Not until many years after the first world war, after Felix Klein had gone and Richard Courant had succeeded him, towards the end of the brief period of the German Republic, did Klein's dream of the Mathematical Institute at Göttingen come true. But soon the Nazi storm broke and those who had laid the plans and who taught there with Hilbert were scattered over the earth, and the years after 1933 became for him tragic years of ever deepening loneliness.

Hilbert was of slight build. Above the small lower face with its goatee there rose the dome of a powerful, in later years bald, skull. He was physically agile, a tireless walker, a good skater, and a passionate gardener. Until 1925 he enjoyed good health. Then he fell ill of pernicious anaemia. Yet this illness only temporarily paralyzed his restless activity in teaching and research. He was among the first with whom the liver treatment, inaugurated by G. R. Minot at Harvard, proved successful; undoubtedly it saved Hilbert's life at that time. Hilbert's research left an indelible imprint on practically all branches of mathematical science. Yet in distinct successive periods he devoted himself with impassioned exclusiveness to but a single subject at a time. Perhaps his deepest investigations are those on the theory of number fields. His monumental report on the "Theorie der algebraischen Zahlkörper" ["Theory of Algebraic Number Fields"] which he submitted to the Deutsche Mathematiker-Vereinigung, is dated as of the year 1897, and as far as I know Hilbert did not publish another paper in this field after 1899.[2] The methodical unity of mathematics was for him a matter of belief and experience. It appeared to him essential that — in the face of the manifold interrelations and for the sake of the fertility of research — the productive mathematician should make himself at home in all fields. To quote his own words: "The question is forced upon us whether mathematics is once to face what other sciences have long ago experienced, namely to fall apart into subdivisions whose representatives are hardly able to understand each other and whose connexions for this reason will become ever looser. I neither believe nor wish this to happen; the science of mathematics as I see it is an indivisible whole, an organism whose ability to survive rests on the connexion between its parts." Theoretical physics also was drawn by Hilbert into the domain of his research; during a whole decade beginning in 1912 it stood at the centre of his interest. Great, fruitful problems appear to him as the life nerve of mathematics. "Just as every human enterprise prosecutes final aims," says he, "so mathematical research needs problems. Their solution steels the force of the investigator." In his famous lecture at the International Congress of Mathematicians at Paris in 1900 Hilbert tries to probe the immediate future of mathematics by posing twenty-three unsolved problems; they have indeed, as we can state today in retrospect, played an eminent role in the development of mathematics during the subsequent forty-three years. A characteristic feature of Hilbert's method is a peculiarly direct attack on problems, unfettered by any algorithms; he always goes back to the questions in their original simplicity. When it is a matter of transferring the theory of linear equations from a finite to an infinite number of unknowns he begins by getting rid of the calculatory tool of determinants. A truly great example of far-reaching significance is his mastery of Dirichlet's principle which, originally springing from mathematical physics, provided Riemann with the foundation of his theory of algebraic functions and Abelian integrals, but which subsequently had fallen a victim of Weierstrass's pitiless criticism.[3] Hilbert salvaged it in its entirety. The whole finely wrought apparatus of the Calculus of Variations was here consciously set aside. We only need to mention the names R. Courant and M. Morse to indicate what role this direct method of the calculus of variations was destined to play in recent times. It seems to me that with Hilbert the mastering of single concrete problems and the forming of general abstract concepts are balanced in a particularly fortunate manner. He came of a time in which the algorithm had played a more extensive part, and therefore he stressed strongly a conceptual procedure; but in the meantime our advance in this direction has been so uninhibited and with so little concern for a growth of the problematics in depth that many of us have begun to fear for the mathematical substance. In Hilbert simplicity and rigor go hand in hand. The growing demand for rigor, imposed by the critical reflections of the nineteenth century upon those parts of mathematics which operate in the continuum, was felt by most investigators as a heavy yoke that made their steps dragging and awkward. Full of longing and with uneasiness they looked back upon Euler's era of happy-go-lucky analysis. With Hilbert rigor figures no longer as enemy, but as promoter of simplicity. Yet the secret of Hilbert's creative force is not plumbed by any of these remarks. A further element of it, I feel, was his sensitivity in registering hints which revealed to him general relations while solving special problems This is most magnificently exemplified by the way in which, during his theory of numbers period, he was led to the enunciation of his general theorems on class fields and the general law of reciprocity.

In a few words we shall now recall Hilbert's most important achievements. In the years 1888–1892 he proved the fundamental finiteness theorems of the *theory of invariants* for the full projective group. His method, though yielding a proof for the existence of a finite basis for the invariants, does not enable one actually to construct it in a concrete individual case. Hence the exclamation by the great algorithmician P. Gordan, at the appearance of Hilbert's paper: "This is not mathematics; this is theology!" It reveals an antithesis which reaches down to the very roots of mathematics. Hilbert, however, in further penetrating investigations, furnished the means for a finite execution of the construction.

His papers on the theory of invariants had the unexpected effect of withering, as it were overnight, a discipline which so far had stood in full bloom. Its central problems he had finished once and for all. Entirely different was his effect on the *theory of number fields*, which he took up in the years 1892–1898. It is a great pleasure to watch how, step by step, in a succession of papers ascending from the particular to the general, the adequate concepts and methods are evolved and the essential connexions come to light. These papers proved of extraordinary fertility for the future. On the pure theory of numbers side I mention the names of Furtwängler, Takagi, Artin, Hasse, Chevalley, and on the number- and-function theoretical one, those of Fuëter and Hecke.

During the subsequent period, 1898–1902, the *foundations of geometry* are nearest to Hilbert's heart, and he is seized by the idea of axiomatics. The soil was well prepared, especially by the Italian school of geometers. Yet it was as if over a landscape, wherein but a few men with a superb sense of orientation had found their way in murky twilight, the sun had risen all at once. Clear and clean-cut we find stated the axiomatic concept according to which geometry is a hypothetical deductive system; it depends on the "implicit definitions" of the concepts of spatial objects and relations which the axioms contain, and not on a description of their intuitive content. A complete and natural system of geometric axioms is set up. They are required to satisfy the logical demands of consistency, independence, and completeness, and by means of quite a few peculiar geometries, constructed ad hoc, the proof of independence is furnished in detail. The general ideas appear to us today almost banal, but in these examples Hilbert unfolds his typical wealth of invention. While in this fashion the geometric concepts become formalized, the logical ones function as before in their intuitive significance. The further step where logic too succumbs to formalization, thus giving rise to a purely symbolic mathematics — a step upon which Hilbert already pondered at this epoch, as a paper read to the International Congress of 1904 proves, and which is inevitable for the ultimate justification of the role played by the infinite in mathematics — was systematically followed up by Hilbert during the final years of his mathematical productivity, from 1922 on. In contrast to L. E. J. Brouwer's intuitionism, which finds itself forced to abandon major parts of historical mathematics as untenable, Hilbert attempts to save the holdings of mathematics in their entirety by proving its formalism free of contradiction. Admittedly the question of truth is thus shifted into the question of consistency. To a limited extent the latter has been established by Hilbert himself in collaboration with P. Bernays, by J. von Neumann, and G. Gentzen. In recent times, however, the entire enterprise has become questionable on account of K. Gödel's surprising discoveries. While Brouwer has made clear to us to what extent the intuitively certain falls short of the mathematically provable, Gödel shows conversely to what extent the intuitively certain goes beyond what (in an arbitrary but fixed formalism) is capable of mathematical proof. The question of the ultimate foundations and the ultimate meaning of mathematics remains open; we do not know in which direction it will find its final solution nor even whether a final objective answer can be expected at all. "Mathematizing" may well be a creative activity of man, like language or music, of primary originality, whose historical decisions defy complete objective rationalization.

A chance occasion, a lecture in 1901 by the Swedish mathematician E. Holmgren in Hilbert's seminar dealing with the now classical paper of Fredholm's on *integral equations*, then but recently published, provided the impulse which started Hilbert on his investigations on this subject which absorbed his attention until 1912. Fredholm had limited himself to setting up the analogue of the theory of linear equations, while Hilbert recognized that the analogue of the transformation on to principal axes of quadratic forms yields the theory of the eigenvalues and eigenfunctions for the vibration problems of physics. He developed the parallel between integral equations and sum equations in infinitely many unknowns, and subsequently proceeded to push ahead from the spectral theory of "completely continuous" to the much more general one of "bounded" quadratic forms. Today these things present themselves to us in the framework of a general theory of Hilbert space. Astonishing indeed is the variety of interesting applications which integral equations find in the most diverse branches of mathematics and physics. Mention should be made of Hilbert's own solution of Riemann's problem of monodromy for linear differential equations, a far-reaching generalisation of the existence theorem for algebraic functions on a preassigned Riemann surface, and his treatment of the kinetic theory of gases, also the completeness theorem for the representations of a continuous compact group, and finally in recent times the construction of harmonic integrals on an arbitrary Riemannian manifold, successfully accomplished by the use of Hilbertian means.[4] Thus only under Hilbert's hands did the full fertility of Fredholm's great idea unfold. But it was also due to his influence that the theory of integral equations became a world-wide fad in mathematics for a considerable length of time, producing an enormous literature, for the most part of rather ephemeral value. It was not merit but a favor of fortune when, beginning in 1923 (Heisenberg, Schrödinger) the spectral theory in Hilbert space was discovered to be the adequate mathematical instrument of quantum physics. This later impulse led to a re-examination of the entire complex of problems with refined means (J. von Neumann, M. Stone, and others).

This period of integral equations is followed by Hilbert's physical period. Significant though it was for Hilbert's complete personality as a scientist, it produced a lesser harvest than the purely mathematical phases, and may here be passed over. I shall mention instead two single, somewhat isolated, accomplishments that were to have a great effect: his vindication of Dirichlet's principle; and his proof of a famous century-old conjecture of Waring's, carrying the statement that every integer can be written as a sum of four squares over from squares to arbitrary powers. The physical period is finally succeeded by the last one, already mentioned above, in the course of which Hilbert gives an entirely new turn to the question concerning the foundation and the truth content of mathematics itself. A fruit of Hilbert's pedagogic activity during this period is the charming book by him and Cohn-Vossen, *Anschauliche Geometrie* [*Intuitive Geometry*].[5]

This summary, though far from being complete, may suffice to indicate the universality and depth of Hilbert's mathematical work. He has impressed the seal of his spirit upon a whole era of mathematics. And yet I do not believe that his research work alone accounts for the brilliance that eradiated from him, nor for his tremendous influence. Gauss and Riemann, to mention two other Göttingers, were greater mathematicians than Hilbert, and yet their immediate effect upon their contemporaries was undoubtedly smaller. Part of this is certainly due to the changing conditions of time, but the character of the men is probably more decisive. Hilbert's was a nature filled with the zest of living, seeking the intercourse of other people, and delighting in the exchange of scientific ideas. He had his own free manner of learning and teaching. His comprehensive mathematical knowledge he acquired not so much from lectures as in conversations with Minkowski and Hurwitz. "On innumerable walks, at times undertaken day by day," he tells in his obituary on Hurwitz, "we browsed in the course of eight years through every corner of mathematical science." And as he had learned from Hurwitz, so he taught in later years his own pupils on far-flung walks through the woods surrounding Göttingen or, on rainy days, as peripatetics, in his covered garden walk. His optimism, his spiritual passion, and his unshakable faith in the value of science were irresistibly infectious. He says: "The conviction of the solvability of each and every mathematical problem spurs us on during our work; we hear within ourselves the steady call: there is the problem; search for the solution. You can find it by sheer thinking, for in mathematics there is no *ignorabimus* ["we shall not know"]."[6] His enthusiasm did get along with criticism, but not with scepticism. The snobbish attitude of pretended indifference, of "merely fooling around with things" or even of playful cynicism, did not exist in his circle. Hilbert was enormously industrious; he liked to quote Lichtenberg's saying: "Genius is industry." Yet for all this there was light and laughter around him. Under the influence of his dominating power of suggestion one readily considered important whatever he did; his vision and experience inspired confidence in the fruitfulness of the hints he dropped. It is moreover decisive that he was not merely a scientist but a scientific personality, and therefore capable not only of teaching the technique of his science but also of being a spiritual leader. Although not committing himself to one of the established epistemological or metaphysical doctrines, he was a philosopher in that he was concerned with the life of the idea as it realizes itself among men and as an indivisible whole; he had the force to evoke it, he felt responsible for it in his own sphere, and measured his individual scientific efforts against it. Last, not least, the environment also helped. A university such as Göttingen, in the halcyon days before 1914, was particularly favourable for the development of a living scientific school. Once a band of disciples had gathered around Hilbert, intent upon research and little worried by the toil of teaching, it was but natural that in joint competitive aspiration of related aims each should stimulate the other; there was no need that everything come from the master.

His homeland and America among all countries felt Hibert's impact most thoroughly. His influence upon American mathematics was not restricted to his immediate pupils. Thus, for instance, the Hilbert of the foundations of geometry had a profound effect on E. H. Moore and O. Veblen; the Hilbert of integral equations on George D. Birkhoff.

A picture of Hilbert's personality should also touch upon his attitude regarding the great powers in the lives of men; social and political organization, art, religion, morals and manners, family, friendship, love. Suffice it to say here that he was singularly free from all national and racial prejudices, that in all questions, political, social, or spiritual, he stood forever on the side of freedom, frequently in isolated opposition to the compact majority of his environment. Unforgotten by all those present remains the unanimous and prolonged applause which greeted him in 1928 at Bologna, the first International Congress of Mathematicians at which, following a lengthy struggle, the Germans were once more admitted. It was a telling expression of veneration for the great mathematician whom every one knew to have risen from a severe illness, but at the same time an expression of respect for the independent attitude, *au dessus de la mêlée*, from which he had not wavered during the world conflict. With veneration, gratitude and love his memory will be preserved beyond the gates of death by many a mathematician throughout the world.

Notes

[Published as Weyl 1944a; *WGA* 4:121–129.]

[1] [For these academic ranks, see above, 12, 66.]

[2] [Hilbert's *Zahlbericht* (*Number report*) "was the principal textbook on algebraic number theory for a period of at least thirty years after its appearance" in 1897, according to its English translation, Hilbert 1998, whose introduction (1–15) gives a helpful overview. See also Corry 2004a for a discussion of this work in context.]

[3] [See Gowers et al. 2008, 125–126, 475–476. For Dirichlet's principle, see below, 180.]

[4] [For the Riemann-Hilbert problem, see Its 2003.]

[5] [Available in English as Hilbert and Cohn-Vossen 1999. For Waring's problem, see Vaughn and Wooley 2002.]

[6] [Hilbert is referring to a famous 1880 lecture by Emil du Bois-Reymond, who had argued that "we shall not know" [*ignorabimus*] the answers to a number of transcendent "world riddles," such as the ultimate origins of matter, force, and motion; see du Bois-Reymond 1907, 1974.]

David Hilbert and His Mathematical Work
(1944)

A great master of mathematics passed away when David Hilbert died in Göttingen on February the 14th, 1943, at the age of eighty-one. In retrospect it seems to us that the era of mathematics upon which he impressed the seal of his spirit and which is now sinking below the horizon achieved a more perfect balance than prevailed before and after, between the mastering of single concrete problems and the formation of general abstract concepts. Hilbert's own work contributed not a little to bringing about this happy equilibrium, and the direction in which we have since proceeded can in many instances be traced back to his impulses. No mathematician of equal stature has risen from our generation.

America owes him much. Many young mathematicians from this country, who later played a considerable role in the development of American mathematics, migrated to Göttingen between 1900 and 1914 to study under Hilbert. But the influence of his problems, his viewpoints, his methods, spread far beyond the circle of those who were directly inspired by his teaching.

Hilbert was singularly free from national and racial prejudices; in all public questions, be they political, social or spiritual, he stood forever on the side of freedom, frequently in isolated opposition against the compact majority of his environment. He kept his head clear and was not afraid to swim against the current, even amidst the violent passions aroused by the first world war that swept so many other scientists off their feet. It was not mere chance that when the Nazis "purged" the German universities in 1933 their hand fell most heavily on the Hilbert school and that Hilbert's most intimate collaborators left Germany either voluntarily or under the pressure of Nazi persecution. He himself was too old, and stayed behind; but the years after 1933 became for him years of ever deepening tragic loneliness.

It was another Germany in which he was born on January 23, 1862, and grew up. Königsberg, the eastern outpost of Prussia, the city of Kant, was his home town. Contrary to the habit of most German students who used to wander from university to university, Hilbert studied at home, and it was in his home university that he climbed the first rungs of the academic ladder, becoming *Privatdozent* and in due time *ausserordentlicher Professor*.[1] During his entire life he preserved uncorrupted the characteristic Baltic accent. His reputation as a leading algebraist was well established when on Felix Klein's initiative he was offered a full professorship at Göttingen in 1895. From then on until the end of his life Hilbert remained in Göttingen. He was retired in 1930.

When one inquires into the dominant influences acting upon Hilbert in his formative years, one is puzzled by the peculiarly ambivalent character of his relationship to Kronecker: dependent on him, he rebels against him. Kronecker's work is undoubtedly of paramount importance for Hilbert in his algebraic period. But the old gentleman in Berlin, so it seemed to Hilbert, used his power and authority to stretch mathematics upon the Procrustean bed of arbitrary philosophical principles and to suppress such developments as did not conform: Kronecker insisted that existence theorems should be proved by explicit construction, in terms of integers, while Hilbert was an early champion of Georg Cantor's general set-theoretic ideas. Personal reasons added to the bitter feeling.[2] A late echo of this old feud is the polemic against Brouwer's intuitionism with which the sexagenarian Hilbert opens his first article on "New Foundation of Mathematics" ("Neubegründung der Mathematik," 1922): Hilbert's slashing blows are aimed at Kronecker's ghost whom he sees rising from his grave. But inescapable ambivalence even here — while he fights him he follows him: reasoning along strictly intuitionistic lines is found necessary by him to safeguard non-intuitionistic mathematics.[3]

More decisive than any other influence for the young Hilbert at Königsberg was his friendship with Adolf Hurwitz and Minkowski. He got his thorough mathematical training less from lectures, teachers or books, than from conversation. "During innumerable walks, at times undertaken day after day," writes Hilbert in his obituary on Hurwitz, "we roamed in these eight years through all the corners of mathematical science, and Hurwitz with his extensive, firmly grounded and well-ordered knowledge was for us always the leader." Closer and of a very intimate character was Hilbert's lifelong friendship with Minkowski. The Königsberg circle was broken up when Hurwitz in 1892 left for Zürich, soon to be followed by Minkowski. Hilbert first became Hurwitz's successor in Königsberg and then moved on to Göttingen. The year 1902 saw him and Minkowski reunited in Göttingen where a new chair of mathematics had been created for Minkowski. The two friends became the heroes of the great and brilliant period which our science experienced during the following decade in Göttingen, unforgettable to those who lived through it. Klein, for whom mathematical research had ceased to be the central interest, ruled over it as a distant but benevolent god in the clouds. Too soon was this happy constellation dissolved by Minkowski's sudden death in 1909. In a memorial address before the Göttingen Gesellschaft der Wissenschaften, Hilbert spoke thus about his friend: "Our science, which we loved above everything, had brought us together. It appeared to us as a flowering garden. In this garden there are beaten paths where one may look around at leisure and enjoy oneself without effort, especially at the side of a congenial companion. But we also liked to seek out hidden trails and discovered many a novel view, beautiful to behold, so we thought, and when we pointed them out to one another our joy was perfect."

I quote these words not only as testimony of a friendship of rare depth and fecundity that was based on common scientific interest, but also because I seem to hear in them from afar the sweet flute of the Pied Piper that Hilbert was, seducing so many rats to follow him into the deep river of mathematics. If examples are wanted let me tell my own story. I came to Göttingen as a country lad of eighteen, having chosen that university mainly because the director of my high school happened to be a cousin of Hilbert's and had given me a letter of recommendation to him. In the fullness of my innocence and ignorance, I made bold to take the course Hilbert had announced for that term, on the notion of number and the quadrature of the circle. Most of it went straight over my head. But the doors of a new world swung open for me, and I had not sat long at Hilbert's feet before the resolution formed itself in my young heart that I must by all means read and study whatever this man had written. And after the first year I went home with Hilbert's *Zahlbericht* under my arm, and during the summer vacation I worked my way through it — without any previous knowledge of elementary number theory or Galois theory.[4] These were the happiest months of my life, whose shine, across years burdened with our common share of doubt and failure, still comforts my soul.

The impact of a scientist on his epoch is not directly proportional to the scientific weight of his research. To be sure, Hilbert's mathematical work is of great depth and universality, and yet his tremendous influence is not accounted for by it alone. Gauss and Riemann, to mention two other Göttingers, are certainly of no lesser stature than Hilbert, but they made little stir among their contemporaries and no "school" of devoted followers formed around them. No doubt this is due in part to the changing conditions of time, but the character of the men is probably more decisive. A taste for solitude, even obscurity, is in no way irreconcilable with great creative gifts. But Hilbert's was a nature filled with the zest of living, seeking intercourse with other people, above all with younger scientists, and delighting in the exchange of ideas. Just as he had learned from Hurwitz, so he taught his own pupils, at least those in whom he took a deeper personal interest: on far-flung walks through the woods surrounding Göttingen or, on rainy days, as "peripatetics" in his covered garden walk. His optimism, his spiritual passion, his unshakable faith in the supreme value of science, and his firm confidence in the power of reason to find simple and clear answers to simple and clear questions were irresistibly contagious. If Kant through critique and analysis arrived at the principle of the supremacy of practical reason, Hilbert incorporated, as it were, the supremacy of pure reason — sometimes with laughing arrogance (*arrogancia* in the Spanish sense), sometimes with the ingratiating smile of intellect's spoiled child, but most of the time with the seriousness of a man who believes and must believe in what is the essence of his own life. His enthusiasm was compatible with critical acumen, but not with scepticism. The snobbish attitude of pretended indifference, of "merely fooling around with things," or even of playful cynicism, were unknown in his circle. You had better think twice before you uttered a lie or an empty phrase to him: his directness could be something to be afraid of. He was enormously industrious and liked to quote Lichtenberg's saying: "Genius is industry." Yet for all this there was light and laughter around him. He had great suggestive power; it sometimes lifted even mediocre minds high above their natural level to astonishing, though isolated achievements. I do not remember which mathematician once said to him: "You have forced us all to consider important those problems which you considered important." His vision and experience inspired confidence in the fruitfulness of the hints he dropped. He did not hide his light under a bushel. In his papers one encounters not infrequently utterances of pride in a beautiful or unexpected result, and in his legitimate satisfaction he sometimes did not give to his predecessors on whose ideas he built all the credit they deserved. The problems of mathematics are not isolated problems in a vacuum; there pulses in them the life of ideas which realize themselves *in concreto* through our human endeavors in our historical existence, but forming an indissoluble whole transcend any particular science. Hilbert had the power to evoke this life; against it he measured his individual scientific efforts and felt responsible for it in his own sphere. In this sense he was a philosopher, not in the sense of adhering to one of the established epistemological or metaphysical doctrines. Does not in such personal qualities of the academic teacher, rather than in any objectivities or universally accepted metaphysics, lie the answer to Hutchins's quest for a true *universitas literarum*?[5]

Were it my aim to give a full picture of Hilbert's personality I should have to touch upon his attitude regarding the great powers in the lives of men: social and political organization, art, religion, morals and manners, family, friendship, love, and I should also probably have to indicate some of the shadows cast by so much light. I wanted merely to sketch the mathematical side of his personality in an attempt to explain, however incompletely, the peculiar charm and the enormous influence which he exerted. In appraising the latter one must not overlook the environmental factor. A German university in a small town like Göttingen, especially in the halcyon days before 1914, was a favorable milieu for the development of a scientific school. The high social prestige of the professors and everything connected with the university created an atmosphere the like of which has hardly ever existed in America. And once a band of disciples had gathered around Hilbert, intent on research and little worried by the chore of teaching, how could they fail to stimulate one another! We have seen the same thing happening here at Princeton during the first years of the Institute for Advanced Study; there is a kind of snowball effect in the formation of such condensation points of scientific research.

Before giving a more detailed account of Hilbert's work, it remains to characterize in a few words the peculiarly Hilbertian brand of mathematical thinking. It is reflected in his literary style which is one of great *lucidity*. It is as if you were on a swift walk through a sunny open landscape; you look freely around, demarcation lines and connecting roads are pointed out to you, before you must brace yourself to climb the hill; then the path goes straight up, no ambling around, no detours. His style has not the terseness of many of our modern authors in mathematics, which is based on the assumption that printer's labor and paper are costly but the reader's effort and time are not. In carrying out a complete induction Hilbert finds time to develop the first two steps before formulating the general conclusion from n to $n + 1$. How many examples illustrate the fundamental theorems of his algebraic papers — examples not constructed ad hoc, but genuine ones worth being studied for their own sake!

In Hilbert's approach to mathematics, *simplicity* and *rigor* go hand in hand. The generation before him, nay even most analysts of his time, felt the growing demand for rigor imposed upon analysis by the critique of the nineteenth century, which culminated in Weierstrass, as a heavy yoke that made their steps dragging and awkward. Hilbert did much to change that attitude. In his famous address, *Mathematische Probleme*, delivered before the Paris Congress in 1900, he stresses the importance of great concrete fruitful *problems*.[6] "As long as a branch of science," says he, "affords an abundance of problems, it is full of life; want of problems means death or cessation of independent development. Just as every human enterprise prosecutes final aims, so mathematical research needs problems. Their solution steels the force of the investigator; thus he discovers new methods and viewpoints and widens his horizon." "One who without a definite problem before his eyes searches for methods, will probably search in vain." The *methodical unity of mathematics* was for him a matter of belief and experience. Again I quote his own words: "The question is forced upon us whether mathematics is once to face what other sciences long ago experienced, namely the falling apart into subdivisions whose representatives are hardly able to understand each other and whose connections for this reason will become ever looser. I neither wish nor believe it. The science of mathematics as I see it is an indivisible whole, an organism whose ability to survive rests on the connection between its parts." A characteristic feature of Hilbert's method is a peculiarly *direct attack* on problems, unfettered by algorithms; he always goes back to the questions in their original simplicity. An outstanding example is his salvage of Dirichlet's principle which had fallen a victim of Weierstrass's criticism, but his work abounds in similar examples. His strength, equally disdainful of the convulsion of Herculean efforts and of surprising tricks and ruses, is combined with an uncompromising *purity*.[7]

Hilbert helped the reviewer of his work greatly by seeing to it that it is rather neatly cut into different periods during each of which he was almost exclusively occupied with one particular set of problems. If he was engrossed in integral equations, integral equations seemed everything; dropping a subject, he dropped it for good and turned to something else. It was in this characteristic way that he achieved universality. I discern five main periods: i. Theory of invariants

(1885–1893). ii. Theory of algebraic number fields (1893–1898). iii. Foundations, (a) of geometry (1898–1902), (b) of mathematics in general (1922–1930). iv. Integral equations (1902–1912). v. Physics (1910–1922). The headings are a little more specific than they ought to be. Not all of Hilbert's algebraic achievements are directly related to invariants. His papers on calculus of variations are here lumped together with those on integral equations. And of course there are some overlappings and a few stray children who break the rules of time, the most astonishing his proof of Waring's theorem in 1909.

His Paris address on "Mathematical problems" quoted above straddles all fields of our science. Trying to unveil what the future would hold in store for us, he posed and discussed twenty-three unsolved problems which have indeed, as we can now state in retrospect, played an important role during the following forty odd years. A mathematician who had solved one of them thereby passed on to the honors class of the mathematical community.

Literature

Hilbert's *Gesammelte Abhandlungen* were published in 3 volumes by J. Springer, Berlin, 1932–35 [Hilbert 1932–35]. This edition contains his *Zahlbericht,* but not his two books:

Grundlagen der Geometrie, 7th ed., Leipzig, 1930. [Hilbert 1971]

Grundzüge einer allgemeinen Theorie der linearen Integralgleichungen, Leipzig and Berlin, 1912. [Hilbert 1953]

Hilbert is co-author of the following works :

R. Courant and D. Hilbert, *Methoden der mathematischen Physik,* Berlin, vol. 1, 2nd ed., 1931, vol. 2, 1937. [Courant and Hilbert 1989]

D. Hilbert and W. Ackermann, *Grundzüge der theoretischen Logik,* Berlin, 1928. [Hilbert and Ackermann 1950]

D. Hilbert and S. Cohn-Vossen, *Anschauliche Geometrie,* Berlin, 1932. [Hilbert and Cohn-Vossen 1999]

D. Hilbert and P. Bernays, *Grundlagen der Mathematik,* Berlin, vol. 1,1934, vol. 2, 1939. [Hilbert and Bernays 1968–1970]

The *Collected Papers* contain articles by B. L. van der Waerden, H. Hasse, A. Schmidt, P. Bernays, and E. Hellinger, on Hilbert's work in algebra, in number theory, on the foundations of geometry and arithmetics, and on integral equations. These articles trace the development after Hilbert, giving ample references. The reader may also consult a number of *Die Naturwissenschaften,* vol. 10 (1922), pp. 65–104, dedicated to Hilbert, which surveys his work prior to 1922, and an article by L. Bieberbach, "Ueber den Einfluss von Hilberts Pariser Vortrag über 'Mathematische Probleme' auf die Entwicklung der Mathematik in den letzten dreissig Jahren," *Die Naturwissenschaften,* vol. 18 (1930), pp. 1101–1111. O. Blumenthal wrote a life of Hilbert (*Collected Papers,* vol. 3, pp. 388–429).

I omit all quotations of literature covered by these articles.

Theory of invariants

The classical theory of invariants deals with polynomials $J = J(x_1, \cdots, x_n)$ depending on the coefficients $(x_1, \cdots, x_n)$ of one or several ground forms of a given number of arguments $\eta_1, \cdots, \eta_g$. Any linear substitution s of determinant 1 of the g arguments induces a certain linear transformation $U(s)$ of the variable coefficients $x_1, \cdots, x_n, x \to x' = U(s)x$, whereby $J = J(x_1 \cdots x_n)$ changes into a new form $J(x'_1 \cdots x'_n) = J^s(x_1 \cdots x_n)$. J is an invariant if $J^* = J$ for every s.[8] (The restriction to *unimodular* transformations s enables one to avoid the more involved concept of *relative* invariants and to remove the restriction to homogeneous polynomials, with the convenient consequence that the invariants form a *ring*.) The classical problem is a special case of the general problem of invariants in which s ranges over an arbitrary given abstract group Γ and $s \to U(s)$ is any representation of that group (that is, a law according to which every element s of Γ induces a linear transformation $U(s)$ of the n variables $x_1, \cdots, x_n$ such that the composition of group elements is reflected in composition of the induced transformations). The development before Hilbert had led up to two main theorems, which however had been proved in very special cases only. The first states that the invariants have a finite integrity basis, or that we can pick a finite number among them, say $i_1, \cdots, i_m$, such that every invariant J is expressible as a polynomial in $i_1, \cdots, i_m$. An identical relation between the basic invariants $i_1, \cdots, i_m$ is a polynomial $F(z_1 \cdots z_m)$ of m independent variables $z_1, \cdots z_m$ which vanishes identically by virtue of the substitution

$$z_1 = i_1(x_1 \cdots x_n), \cdots, z_m = i_m(x_1 \cdots x_n).$$

The second main theorem asserts that *the relations have a finite ideal basis*, or that one can pick a finite number among them, say $F_1, \cdots, F_h$, such that every relation F is expressible in the form

$$(1) \qquad\qquad F = Q_1 F_1 + \cdots + Q_h F_h,$$

the Q_i being polynomials of the variables $z_1, \cdots, z_m$.

I venture the guess that Hilbert first succeeded in proving the *second* theorem. The relations F form a subset within the ring $k[z_1 \cdots z_m]$ of all polynomials of $z_1, \cdots, z_m$ the coefficients of which lie in a given field k. When Hilbert found his simple proof he could not fail to notice that it applied to any set of polynomials Σ whatsoever and he thus discovered one of the most fundamental theorems of algebra, which was instrumental in ushering in our modern abstract approach, namely:

(A) Every subset S of the polynomial ring $k[z_1 \cdots z_m]$ has a finite ideal basis.

Is it bad metaphysics to add that his proof turned out so simple *because* the proposition holds in this generality? The proof proceeds by the adjoining of one variable z_i after the other, the individual step being taken care of by the statement: If a given *ring* r satisfies the condition (P): that every subset of r has a finite ideal basis, then the ring $r[z]$ of polynomials of a single variable$z$ with coefficients in r satisfies the same condition (P). Once this is established one gets not only (A) but also an arithmetic refinement discussed by Hilbert in which the field k of rational numbers is replaced by the ring of rational integers.

The subset Σ of relations to which Hilbert applies his theorem (A) is itself an *ideal,* and thus the ideal $\{F_1, \cdots, F_h\}$, that is, the totality of all elements of the form (1), $Q_i \in k[z_i \cdots z_m]$, not only contains, but coincides with, Σ. The proof, however, works even if Σ is not an ideal, and yields at one stroke (1) the enveloping ideal $\{\Sigma\}$ of Σ and (2) the reduction of that ideal to a finite basis, $\{\Sigma\} = \{F_1 \cdots, F_h\}$.

Construction of a full set of relations $F_1 \cdots, F_h$ would finish the investigation of the algebraic structure of the ring of invariants were it true that any relation F can be represented in the form (1) *in one way only.* But since, generally speaking, this is not so, we must ask for the "vectors of polynomials" $M = (M_1, \cdots, M_h)$ for which $M_1 F_1 + \cdots + M_h F_h$ vanishes identically in z (syzygy of first order). These linear relations M between $F_1 \cdots, F_h$ again form an ideal to which Theorem (A) is applicable, the basis of the M thus obtained giving rise to syzygies of the second order. To the first two main theorems Hilbert adds a third to the effect that if redundance is avoided, the chain of syzygies breaks off after at most m steps.

All this hangs in the air unless we can establish the *first main theorem*, which is of an altogether different character because it asks for an integrity, not an ideal basis. Discussing invariants we operate in the ring $k_x = k[x_i \cdots x_n]$ of polynomials of $x_i \cdots x_n$ in a given field k. Hilbert applies his Theorem (A) to the totality $\mathfrak{J}$ of all *invariants J for which* $J(0, \cdots, 0) = 0$ (a subring of k_x which, by the way, is not an ideal!) and thus determines an ideal basis $i_1, \cdots, i_m$ of $\mathfrak{J}$. Each of the invariants $i = i_r$ may be decomposed into a sum $i = i^{(1)} + i^{(2)} + \cdots$ of homogeneous forms of degree $1, 2, \cdots$, and as the summands are themselves invariants, the i_r may be assumed to be homogeneous forms of degrees $\nu_r \geq 1$. Hilbert then claims that the $i_1, \cdots, i_m$ constitute an integrity basis for all invariants. I use a finite group Γ consisting of N elements s (although this case of the general problem of invariants was never envisaged by Hilbert himself) in order to illustrate the idea by which the transition is made. Every invariant J is representable in the form

$$(2) \qquad\qquad J = c + L_1 i_1 + \cdots + L_m i_m \qquad (L_r \in k_x)$$

where c is the constant $J(0)$. If J is of degree ν one may lop off in L_r all terms of higher degree than $\nu - \nu_r$ without destroying the equation. If it were possible by some process to change the coefficients L in (2) into invariants, the desired result would follow by induction with respect to the degree of J. In the case of a finite group such a process is readily found: the process of averaging. The linear transformation $U(s)$ of the variables $x_i, \cdots, x_n$ induced by s carries (2) into

$$J = c + L_1^s \cdot i_1 + \cdots + L_m^s \cdot i_m.$$

Summation over s and subsequent division by the number N yields the relation

$$J = c + L_1^* i_1 + \cdots + L_m^* i_m$$

where

$$L_r^* = \frac{1}{N} \cdot \sum_s L_r^s.$$

It is of the same nature as (2), except for the decisive fact that according to their formation the new coefficients L^* are invariants.[9]

Actually Hilbert had to do, not with a finite group but with the classical problem in which the group Γ consists of all linear transformations s of g variables $\eta_1, \cdots, \eta_g$, and instead of the averaging process he had to resort to a differentiation process invented by Cayley, the so-called Cayley Ω-process, which he skillfully adapted for his end. (It is essential in Cayley's process that the g^2 components of the matrix s are independent variables; instead of the absolutely invariant polynomials J one has to consider relatively invariant homogeneous *forms* each of which has a definite degree and weight.)

Hilbert's theorem (A) is the foundation stone of the general theory of *algebraic manifolds*. Let us now think of k more specifically as the field of all complex numbers. It seems natural to define an algebraic manifold in the space of n coordinates $x_1, \cdots, x_n$ by a number of simultaneous algebraic equations $F_1 = 0, \cdots, f_h = 0$ ($f_i \in k_x$). According to Theorem (A), nothing would be gained by admitting an infinite number of equations. Let us denote by $Z(f_1, \cdots, f_h)$ the set of points $x = (x_1, \cdots, x_n)$ where $f_1, \cdots, f_h$ and hence all elements of the ideal $\mathcal{J} = \{f_1, \cdots, f_h\}$ vanish simultaneously. g vanishes on $Z(f_1, \cdots, f_h)$ whenever $g \in \{f_1, \cdots, f_h\}$, but the converse is not generally true. For instance x_1 vanishes wherever x_1^3 does, and yet x_1 is not of the form $x_1^3 \cdot q(x_1 \cdots x_n)$. The language of the algebraic geometers distinguishes here between the simple plane $x_1 = 0$ and the triple plane, although the point set is the same in both cases. Hence what they actually mean by an algebraic manifold is the polynomial ideal and not the point set of its zeros. But even if one cannot expect that every polynomial g vanishing on $Z(f_1, \cdots, f_h) = Z(\mathcal{J})$ is contained in the ideal $\mathcal{J} = \{f_1, \cdots, f_h\}$ one hopes that at least some *power* of g will be. Hilbert's *"Nullstellensatz"* ["zero-locus-theorem"] states that this is true, at least if k is the field of complex numbers. It holds in an arbitrary coefficient field k provided one admits points x the coordinates x_i of which are taken from k *or any algebraic extension of* k. Clearly this *Nullstellensatz* goes to the root of the very concept of algebraic manifolds.[10]

Actually Hilbert conceived it as a tool for the investigation of invariants. As we are now dealing with the full linear group let us consider homogeneous invariants only and drop the adjective homogeneous. Exclude the constants (the invariants of degree 0). Suppose we have ascertained μ non-constant invariants $J_1, \cdots, J_\mu$ such that every non-constant invariant vanishes wherever they vanish simultaneously. An ideal basis of the set $\mathfrak{J}$ of all non-constant invariants certainly meets the demand, but a system $J_1, \cdots, J_\mu$ may be had much more cheaply. Indeed, by a beautiful combination of ideas Hilbert proves that if for a given point $x = x^0$ there exists at all an invariant which neither vanishes for $x = x^0$ nor reduces to a constant, then there exists such an invariant whose weight does not exceed a certain a priori limit W (for example, $W = 9n(3n+l)^8$ for a ternary ground form of order n). Hence the $J_1, \cdots, J_\mu$ may be chosen from the invariants of weight not greater than W, and they thus come within the grasp of explicit algebraic construction.

When Hilbert published his proof for the existence of a finite ideal basis, Gordan the formalist, at that time looked upon as the king of invariants, cried out: "This is not mathematics, it is theology!" Hilbert remonstrated then, as he did all his life, against the disparagement of existential arguments as "theology," but we see how, by digging deeper, he was able to meet Gordan's constructive demands. By combining the *Nullstellensatz* with the Cayley process he further showed that every invariant J is an integral *algebraic* (though not an integral *rational*) function of $J_1, \cdots, J_\mu$, satisfying an equation

$$J^e + G_1 J^{e-1} + \cdots + G_e = 0$$

in which the G's are polynomials of $J_1, \cdots, J_\mu$. Hence it must be possible by suitable algebraic extensions to pass from $J_1, \cdots, J_\mu$ to a full integrity basis. From there on familiar algebraic patterns such as were developed by Kronecker and as are amenable to explicit construction may be followed.

After the formal investigations from Cayley and Sylvester to Gordan, Hilbert inaugurated a new epoch in the theory of invariants. Indeed, by discovering new ideas and introducing new powerful methods he not only brought the subject up to the new level set for algebra by Kronecker and Dedekind, but made such a thorough job of it that he all but finished it, at least as far as the full linear group is concerned. With justifiable pride he concludes his paper, *Ueber die vollen In-variantensysteme* ["On Complete Systems of Invariants"] with the words: "Thus I believe the most important goals of the theory of the function fields generated by invariants have been attained," and therewith quits the scene.[11]

Of later developments which took place after Hilbert quit, two main lines seem to me the most important: (1) The averaging process, which we applied above to finite groups, carries over to continuous compact groups. By this transcendental process of integration over the group manifold, Adolf Hurwitz treated the real orthogonal group. The method has been of great fertility. The simple remark that invariants for the real orthogonal group are *eo ipso* also invariant under the full complex orthogonal group indicates how the results can be transferred even to non-compact groups, in particular, as it turns out, to all semi-simple Lie groups. (2) Today the theory of invariants for arbitrary groups has taken its natural place within the frame of the theory of representations of groups by linear substitutions, a development which owes its greatest impulse to G. Frobenius.

Although the first main theorem has been proved for wide classes of groups Γ we do not yet know whether it holds for every group. Such attempts as have been made to establish it in this generality were soon discovered to have failed. A promising line for an algebraic attack is outlined in item 14 of Hilbert's Paris list of Mathematical Problems.

Having dwelt in such detail on Hilbert's theory of invariants, we must be brief with regard to his other, more isolated, contributions to algebra. The first paper in which the young algebraist showed his real mettle concerns the conditions under which a form with real coefficients is representable as a sum of squares of such forms, in particular with the question whether the obviously necessary condition that the form be positive for all real values of its arguments is sufficient. By ingenious continuity arguments and algebraic constructions Hilbert finds three special cases for which the answer is affirmative, among them of course the positive definite quadratic forms, but counterexamples for all other cases. Similar methods recur in two papers dealing with the attractive problem of the maximum number of real ovals of an algebraic curve or surface and their mutual position. Hilbert conjectured that, irrespective of the number of variables, every *rational function* with real (or rational) coefficients is a sum of squares of such functions provided its values are positive for real values of the arguments; and in his *Grundlagen der Geometrie* [*Foundations of Geometry*, Hilbert 1971] he pointed out the role of this fact for the geometric constructions with ruler and "*Eichmass* [gauge or compass]." Later O. Veblen conceived, as the basis of the distinction between positive and negative in any field, the axiom that no square sum equals zero. Independently of him, E. Artin and O. Schreier developed a detailed theory of such "real fields," and by means of it Artin succeeded in proving Hilbert's conjecture.[12]

In passing I mention Hilbert's irreducibility theorem according to which one may substitute in an irreducible polynomial suitable integers for all of the variables but one without destroying the irreducibility of the polynomial, and his paper on the solution of the equation of ninth degree by functions with a minimum number of arguments. They became points of departure for much recent algebraic work (E. Noether, N. Tschebotareff and others). Finally, it ought to be recorded that on the foundations laid by Hilbert a detailed theory of polynomial ideals was erected by E. Lasker and F. S. Macaulay which in turn gave rise to Emmy Noether's general axiomatic theory of ideals. Thus in the field of algebra, as in all other fields, Hilbert's conceptions proved of great consequence for the further development.

Algebraic number fields

When Hilbert, after finishing off the invariants, turned to the theory of *algebraic number fields*, the ground had been laid by Dirichlet's analysis of the group of units more than forty years before, and by Kummer's, Dedekind's and Kronecker's introduction of ideal divisors. The theory deals with an algebraic field κ over the field k of rational numbers. One of the most important general results beyond the foundations had been discovered by Dedekind, who showed that the rational prime divisors of the discriminant of κ are at the same time those primes whose ideal prime factors in κ are not all distinct (ramified primes). l being a rational prime, the adjunction to κ of the lth root of a number α in κ yields a relative cyclic field $K = \kappa(\alpha^{1/l})$ of degree l over κ, *provided κ contains the lth root of unity $\zeta = e^{2\pi i/l}$* (and according to Lagrange, the most general relative cyclic field of degree l over κ is obtained in this fashion). It may be said that it was this circumstance which forced Kummer, as he tried to prove Fermat's theorem of the impossibility of the equation $\alpha^l + \beta^l = \gamma^l$, to pass from the rational ground field k to the cyclotomic field $\kappa_l = k(\zeta)$ and then to conceive his ideal numbers in κ_l and to investigate whether the number of classes of equivalent ideal numbers in κ_l is prime to l. Hilbert sharpened his tools in resuming Kummer's study of the relative cyclic fields of degree l over κ_l which he christened "Kummer fields."

His own first important contribution was a theory of *relative Galois fields K* over a given algebraic number field κ. His main concern is the relation of the Galois group Γ of K/κ to the way in which the prime ideals of κ decompose in K. Given a prime ideal $\mathfrak{P}$ in K of relative degree f, those substitutions s of Γ for which $s\mathfrak{P} = \mathfrak{P}$ form the splitting group. As always in Galois theory one constructs the corresponding subfield of K/κ (splitting field), to which a number of K belongs if it is invariant under all substitutions of the splitting group. The substitutions t which carry every integer $\mathbf{A}$ in K into one, $t\mathbf{A}$, that is congruent to $\mathbf{A}$ mod $\mathfrak{P}$ form an invariant subgroup of of index f, called the inertial group, and the corresponding field (inertial field) is sandwiched in between the splitting field and K. Let $\mathfrak{p}$ be the prime ideal in κ into which $\mathfrak{P}$ goes, and $\mathfrak{P}^e$ the exact power of $\mathfrak{P}$ by which $\mathfrak{p}$ is divisible. I indicate the nature of Hilbert's results by the following central theorem of his: In the splitting field of $\mathfrak{P}$ the prime ideal $\mathfrak{p}$ in κ splits off the prime factor $\mathfrak{p}^* = \mathfrak{P}^e$ of degree 1 (therefore the name!); in passing from the splitting to the inertial field $\mathfrak{p}^*$ stays prime but its degree increases to f; in passing from the inertial to the full field K, $\mathfrak{p}^*$ breaks up into e equal prime factors $\mathfrak{P}$ of the same degree f. For later application I add the following remarks. If $\mathfrak{P}$ goes into $\mathfrak{p}$ in the first power only, $e = 1$ (which is necessarily so provided $\mathfrak{p}$ is not a divisor of the relative discriminant of K/κ) , then the inertial group consists of the identity only. In that case the theory of Galois's strictly finite fields shows that the splitting group is cyclic of order f and that its elements $1, s, s^2, \cdots, s^{f-1}$ are uniquely determined by the congruences

$$sA \equiv A^P, \qquad s^2 A \equiv A^{P^2}, \cdots \quad (\mathrm{mod}\ \mathfrak{P})$$

holding for every integer $\mathbf{A}$. Here P is the number of residues in κ modulo $\mathfrak{p}$ and thus P^f the number of residues in K modulo $\mathfrak{P}$. Today we call $s = \sigma(\mathfrak{P})$ the Frobenius substitution of $\mathfrak{P}$; it is of paramount importance that one particular generating substitution of the splitting group may thus be distinguished among all others. One readily sees that for any substitution u of the Galois group $\sigma(u\mathfrak{P}) = u^{-1} \cdot \sigma(\mathfrak{P}) \cdot u$. Thus if the Galois field K/κ is *Abelian*, the substitution $\sigma(\mathfrak{P}) = \sigma(u\mathfrak{P})$ depends on $\mathfrak{p}$ only and may be denoted by $\left(\frac{K}{\mathfrak{p}}\right)$.

In 1893 the Deutsche Mathematiker-Vereinigung asked Hilbert and Minkowski to submit within two years a report on number theory. Minkowski dropped out after a while. Hilbert's monumental report *Die Theorie der algebraischen Zahlkörper* [*Theory of Algebraic Number Fields*, Hilbert 1998] appeared in the *Jahresberichte* of 1896 (the preface is dated April 1897). What Hilbert accomplished is infinitely more than the Vereinigung could have expected. Indeed, his report is a jewel of mathematical literature. Even today, after almost fifty years, a study of this book is indispensable for anybody who wishes to master the theory of algebraic numbers. Filling the gaps by a number of original investigations, Hilbert welded the theory into an imposing unified body. The proofs of all known theorems he weighed carefully before he decided in favor of those "the principles of which are capable of generalization and the most useful for further research." But before such a selection could be made that "further research" had to be carried out! Meticulous care was given to the notations, with the result that they have been universally adopted (including, to the American printer's dismay, the German letters for ideals!) He greatly simplified Kummer's theory, which rested on very complicated calculations, and he introduced those concepts and proved a number of those theorems in which we see today the foundations of a general theory of relative Abelian fields. The most important concept is the norm residue symbol, a pivotal theorem on relative cyclic fields, his famous Satz 90 (Collected Works, vol. I, p. 149). From the preface in which he describes the general character of number theory, and the topics covered by his report in particular, let me quote one paragraph :

"The theory of number fields is an edifice of rare beauty and harmony. The most richly executed part of this building, as it appears to me, is the theory of Abelian fields which Kummer by his work on the higher laws of reciprocity, and Kronecker by his investigations on the complex multiplication of elliptic functions, have opened up to us. The deep glimpses into the theory which the work of these two mathematicians affords reveals at the same time that there still lies an abundance of priceless treasures hidden in this domain, beckoning as a rich reward to the explorer who knows the value of such treasures and with love pursues the art to win them."

Hilbert himself was the miner who during the following two years brought to light much of the hidden ore. The analogy with the corresponding problems in the realm of algebraic functions of one variable where Riemann's powerful instruments of topology and Abelian integrals are available was for him a guiding principle throughout (cf. his remarks in item 12 of his Paris Problems). It is a great pleasure to watch how, step by step, advancing from the special to the general, Hilbert evolves the adequate concepts and methods, and the essential conclusions emerge. I mention his great paper dealing with the relative quadratic fields, and his last and most important *Ueber die Theorie der relativ Abelschen Zahlkörper* ["On the Theory of Relative Abelian Number Fields"]. On the basis of the examples he carried through in detail, he conceived as by divination and formulated the basic facts about the so-called class fields. Whereas Hilbert's work on invariants was an end, his work on algebraic numbers was a beginning. Most of the labor of such number theorists of the last decades, as Furtwängler, Takagi, Hasse, Artin, Chevalley, has been devoted to proving the results anticipated by Hilbert. By deriving from the ζ-function the existence of certain auxiliary prime ideals, Hilbert had leaned heavily on transcendental arguments. The subsequent development has gradually eliminated these transcendental methods and shown that though they are the fitting and powerful tool for the investigation of the distribution of prime ideals they are alien to the problem of class fields. In attempting to describe the main issues I shall not ignore the progress and simplification due to this later development.

Hilbert's theory of norm residues is based on the following discoveries of his own: (1) he conceived the basic idea and defined the norm residue symbol for all non-exceptional prime spots; (2) he realized the necessity of introducing infinite prime spots; (3) he formulated the general law of reciprocity in terms of the norm symbol; (4) he saw that by means of that law one can extend the definition of the norm symbol to the exceptional prime spots where the really interesting things happen. — It was an essential progress when E. Artin later (5) replaced the roots of unity by the elements of the Galois group as values of the residue symbol. In sketching Hilbert's problems I shall make use of this idea of Artin's and also of the abbreviating language of (6) Hensel's p-adic numbers and (7) Chevalley's *idèles*.[13]

As everybody knows an integer a indivisible by the prime $p \neq 2$ is said to be a quadratic residue if the congruence $x^2 \equiv a \pmod{p}$ is solvable. Gauss introduced the symbol $\left(\frac{a}{p}\right)$ which has the value $+1$ or -1 according to whether a is a quadratic residue or non-residue mod p, and observed that it is a character, $\left(\frac{a}{p}\right) \cdot \left(\frac{a'}{p}\right) = \left(\frac{aa'}{p}\right)$. Indeed, the p residues modulo p — as whose representatives one may take $0, 1, \cdots, (p-1)$ — form a *field*, and after exclusion of 0 a *group* in which the quadratic residues form a subgroup of index 2. Let $K = k(b^{1/2})$ be a quadratic field which arises from the rational ground field k by adjunction of the square root of the rational number b. An integer $a \neq 0$ is called by Hilbert a *p-adic norm* in K if modulo any given power of p it is congruent to the norm of a suitable integer in K. He sets $\left(\frac{a,K}{p}\right) = +1$ if a is p-adic norm, -1 in the opposite case, and finds that this p-adic norm symbol again is a character. The investigation of numbers modulo arbitrarily high powers of p was systematized by K. Hensel in the form of his p-adic numbers, and I repeat Hilbert's definition in this language: "The rational number $a \neq 0$, or more generally the p-adic number $a_p \neq 0$, is a p-adic norm in K if the equation

$$a_p = \text{Nm } (x + yb^{1/2}) = x^2 - by^2$$

has a p-adic solution $x = x_p, y = y_p$; the norm symbol (a_p, K) equals $+1$ or -1 according to whether or not a_p is (p-adic) norm in K."[14] The p-adic numbers form a field $k(p)$ and after exclusion of 0 a multiplicative group G_p in which, according to Hilbert's result, the p-adic norms in K form a subgroup of index 2 or 1. The cyclic nature of the factor group is the salient point. One easily finds that the p-adic squares form a subgroup G_p^2 of index 4 if $p \neq 2$, of index 8 if $p = 2$, and thus the factor group G_p/G_p^2 is not cyclic and could not be described by a single character. Of course every p-adic square is a p-adic norm in K. Both steps, the substitution of K-norms for squares and the passage from the modulus p to arbitrarily high powers of p; the first step amounting to a relaxation, the second to a sharpening of Gauss's condition for quadratic residues; are equally significant for the success of Hilbert's definition.

Every p-adic number $a_p \neq 0$ is of the form $p^h \cdot e_p$ where e_p is a p-adic unit, and thus a_p is of a definite *order* h (at p). An ordinary rational number a coincides with a definite p-adic number $I_p(a) = a_p$. Here I_p symbolizes a homomorphic projection of k into $k(p)$:

$$I_p(a + a') = I_p(a) + I_p(a'), \qquad I_p(aa') = I_p(a) \cdot I_p(a').$$

The character $\left(\frac{a,K}{p}\right)$ is identical with $(I_p(a), K)$.

We come to Hilbert's second discovery: he realized that simple laws will not result unless one adds to the "finite prime spots" p one infinite prime spot q. By definition the q-adic numbers are the real numbers and $I_p(q)$ is the real number with which the rational number a coincides. Hence the real number a_q is a q-adic norm in K if the equation $a_q = x^2 - by^2$ has a solution in real numbers x, y. Clearly if $b > 0$ or K is real, this is the case for every a_q; if however $b < 0$ or K is imaginary, only the positive numbers a_q are q-adic norms. Hence

$$(a_q, K) = 1 \text{ if } K \text{ real}; \qquad (a_q, K) = \text{sgn } a_q \text{ if } K \text{ imaginary.}$$

The fact that the norm symbol is a character is thus much more easily verified for the infinite prime spot than for the finite ones.

Hilbert's third observation is to the effect that Gauss's reciprocity law with its two supplements may be condensed into the elegant formula

$$(3) \qquad \prod_p (I_p(a), K) = \prod_p \left(\frac{a, K}{p}\right) = 1,$$

the product extending over the infinite and all finite prime spots p. There is no difficulty in forming this product because almost all factors (that is, all factors with but a finite number of exceptions) equal unity. Indeed, if the prime p does not go into the discriminant of K, then $(a_p, K) = 1$ for every p-adic unit a_p. Formula (3) is the first real vindication for the norm residue idea, which must have given Hilbert the assurance that the higher reciprocity laws had to be formulated in terms of norm residues.

A given rational number a assigns to every prime spot p a p-adic number $a_p = I_p(a)$. On which features of this assignment does one rely in forming the product (3)? The obvious answer is given by Chevalley's notion of *idèle*: an idèle **a** is a function assigning to every prime spot p a p-adic number $a_p \neq 0$ which is a p-adic unit for almost all prime spots p. The idèles form a multiplicative group J_k. By virtue of the assignment $a_p = I_p(a)$ every rational number $a \neq 0$ gives rise to an idèle, called the *principal idèle a*. With the idèles **a** at our disposal we might as well return to the notation $\left(\frac{a,K}{p}\right)$ for (a_p, K). The formula

$$(4) \qquad \phi_K(\mathbf{a}) = (\mathbf{a}, K) = \prod_p (a_p, K) = \prod_p \left(\frac{\mathbf{a}, K}{p}\right)$$

defines a character ϕ_K, the norm character, in the group J_k of all idèles. The reciprocity law in Hilbert's form (3) maintains that

$$(5) \qquad (\mathbf{a}, K) = 1$$

if **a** is principal. By the very definition of the norm symbol (a_p, K) the same equation holds if **a** is a norm in K, that is, if a_p is a p-adic norm in K for every prime spot p. Two idèles $\mathbf{a}, \mathbf{a}'$ are said to be equivalent, $\mathbf{a} \sim \mathbf{a}'$, if their quotient $\mathbf{a}'\mathbf{a}^{-1}$ is principal. Let us denote by $\mathrm{Nm}J_K$ the group of all idèles which are equivalent to norms in K. Then (5) holds for all idèles of $\mathrm{Nm}J_K$; it would be good to know that it holds for no other idèles, or, in other words, that $\mathrm{Nm}J_K$ is a subgroup of J_k of index 2.

The stage is now reached where the experiences gathered for a quadratic field K over the rational ground field k may be generalized to any relative Abelian field K over a given algebraic number field $\kappa = k(\theta)$. First a word about the infinite prime spots of κ. The defining equation $f(\theta) = 0$, an irreducible equation in k of some degree m, has m distinct roots $\theta', \theta'', \cdots, \theta^{(m)}$ in the continuum of complex numbers. Suppose that r of them are real, say $\theta', \cdots, \theta^{(r)}$. Then each element α of κ has its r real conjugates $\alpha', \cdots, \alpha^{(r)}$, and $\alpha^{(t)}$ arises from α by a homomorphic projection $I^{(t)}$ of κ into the field of all real numbers,

$$\alpha \to \alpha^{(t)} = I^{(t)}(\alpha) \qquad (t = 1, \cdots, r).$$

We therefore speak of r real infinite prime spots $\mathfrak{q}', \cdots, \mathfrak{q}^{(r)}$ with the corresponding projections $I' = I_{\mathfrak{q}'}, \cdots, I^{(r)}$; the fields $\kappa(\mathfrak{q}'), \cdots, \kappa(\mathfrak{q}^{(r)})$ are identical with the field of all real numbers. Thus α is an nth $\mathfrak{q}'$-adic power if the equation $\alpha' = \xi'^n$ has a real solution ξ'. One sees that this imposes a condition only if n is even, and then requires α' to be positive. (In the complex domain the equation is always solvable whether n be even or odd, and that is the reason why we ignore the complex infinite prime spots altogether.)

The finite prime spots are the prime ideals $\mathfrak{p}$ of κ. In studying a Galois field K/κ of relative degree n we first exclude the ramified ideals $\mathfrak{p}$ which go into the relative discriminant of K/κ. An unramified ideal $\mathfrak{p}$ of κ factors in K into a number g of distinct prime ideals $\mathfrak{P}_1 \cdots \mathfrak{P}_g$ of relative degree f, $fg = n$. It is easily seen that a $\mathfrak{p}$-adic number $\alpha_p \neq 0$ is a $\mathfrak{p}$-adic norm in K if and only if its order (at $\mathfrak{p}$) is a multiple of f. In particular, the $\mathfrak{p}$-adic units are norms. Thus we encounter a situation which is essentially simpler than the one taken care of by Gauss's quadratic residue symbol: the norm character of α_p depends only on the order i at $\mathfrak{p}$ of α_p. It is now clear how to proceed: we choose a primitive fth root of unity ζ and define $(\alpha_p, K) = \zeta^i$ if α_p is of order i. This function of $\alpha_p \neq 0$ is a character which assumes the value 1 for the norms and the norms only. But here is the rub: there is no algebraic property distinguishing the several primitive fth roots of unity from one another. Thus the choice of ζ among them remains arbitrary. One could put up with this if one dealt with one prime ideal only. But when one has to take all prime spots simultaneously into account, as is necessary in forming products of the type (4), then the arbitrariness involved in the choice of ζ for each $\mathfrak{p}$ will destroy all hope of obtaining a simple reciprocity law like (5). I shall forego describing the devices by which Eisenstein, Kummer, Hilbert, extricated themselves from this entanglement. By far the best solution was found by Artin: if K/κ is Abelian, then the Frobenius substitution $\left(\frac{K}{\mathfrak{p}}\right)$ is uniquely determined by K and $\mathfrak{p}$ and is an element of order f of the Galois group Γ of K/κ. Let this element of the Galois group replace ζ in our final definition of the $\mathfrak{p}$-adic norm symbol:

$$(6) \qquad (\alpha_p, K) = \left(\frac{\alpha, K}{\mathfrak{p}}\right) = \left(\frac{K}{\mathfrak{p}}\right)^i \text{ if } \alpha_p \text{ is of order } i \text{ at } \mathfrak{p}.$$

We could now form for any idèles $\boldsymbol{\alpha}$ the product

$$\prod_{\mathfrak{p}}(\alpha_p, K) = \prod_{\mathfrak{p}} \left(\frac{\alpha, K}{\mathfrak{p}}\right) = (\boldsymbol{\alpha}, K)$$

extending over all finite and (real) infinite prime spots $\mathfrak{p}$ and formulate the reciprocity law asserting that $(\boldsymbol{\alpha}, K) = 1$ for any principal idèles $\boldsymbol{\alpha}$ — had we not excluded certain exceptional prime spots in our definition of (α_p, K), namely the infinite prime spots and the ramified prime ideals. In the special case he investigated Kummer had succeeded in obtaining the correct value of (α_p, K) for the exceptional $\mathfrak{p}$ by extremely complicated calculations. Hilbert's fourth discovery is a simple and ingenious method of circumventing this formidable obstacle which blocked the road to further progress. Let us first restrict ourselves to idèles $\boldsymbol{\alpha}$ which are nth powers at our exceptional prime spots; in other words, we assume that the equation $\alpha_p = \xi_p^n$ is solvable for the $\mathfrak{p}$-adic values α_p of $\boldsymbol{\alpha}$ at this finite number of prime spots. There is no difficulty in defining $(\boldsymbol{\alpha}, K)$ under this restriction:

$$(\boldsymbol{\alpha}, K) = \prod_{\mathfrak{p}}{}'(\alpha_\mathfrak{p}, K),$$

the product extending, as indicated by the accent, over the non-exceptional prime spots only, for which we know what $(\alpha_\mathfrak{p}, K)$ means. Under the same restriction we prove (with Artin) the reciprocity law

(7) $(\alpha, K) = 1$ if α is principal,

and observe that by its very definition $(\alpha, K) = 1$ if α is norm. We now return to an *arbitrary* idèle α. It is easily shown that there exists an equivalent idèle $\alpha^* \sim \alpha$ which is an nth power at all exceptional prime spots, but of course there will be plenty of them. However, the restricted law of reciprocity insures that

$$(\alpha^*, K) = \prod_{\mathfrak{p}}{}' (\alpha_{\mathfrak{p}}^*, K)$$

has the same value for every one of the α^*'s, and it is this value which we now denote by (α, K). This definition adopted, the reciprocity law (7) and the statement that $(\alpha, K) = 1$ for every norm α follow at once *without restriction*. Thus the reciprocity law itself is made the tool for getting the exceptional prime spots under control!

Once (α, K) is known for every idèle α we can compute (α_p, K) for a given prime spot $\mathfrak{p}$ and a given $\mathfrak{p}$-adic number $\alpha_p \neq 0$ as the value of (α, K) for that "primary" idèle, also denoted by $\alpha_{\mathfrak{p}}$, which equals α_p at $\mathfrak{p}$ and 1 at any other prime spot. (The idèle α is the product of its primary components, $\alpha = \prod_{\mathfrak{p}} \alpha_{\mathfrak{p}}$.) One expects the following two propositions to hold:

I. $(\alpha_{\mathfrak{p}}, K) = 1$ *if and only if $\alpha_{\mathfrak{p}}$ is a $\mathfrak{p}$-adic norm.*

II. *Given a prime ideal* $\mathfrak{p}, (\alpha_{\mathfrak{p}}, K) = 1$ *for every $\mathfrak{p}$-adic unit $\alpha_{\mathfrak{p}}$ if and only if $\mathfrak{p}$ is unramified.*

The direct parts of I and II:

(I_0) $\alpha_{\mathfrak{p}} = $ norm implies $(\alpha_{\mathfrak{p}}, K) = 1$;

(II_0) p unramified implies $(\alpha_{\mathfrak{p}}, K) = 1$ for every $\mathfrak{p}$-adic unit $\alpha_{\mathfrak{p}}$, were settled above. The converse statement of I_0 is trivial for the nonexceptional prime spots; but owing to the indirect definition of the norm symbol for the exceptional prime spots, the proofs of the converse of I_0 for the exceptional spots and of the converse of II_0 are rather intricate. From II we learn that for none of the ramified prime ideals $\mathfrak{p}$ does the norm character of $\alpha_{\mathfrak{p}}$ depend on the order of $\alpha_{\mathfrak{p}}$ only: this simple feature which makes the definition (6) possible is limited to non-ramified $\mathfrak{p}$. One would also expect:

III. *If the principal idèle α is an idèle norm in K, then the number α is norm of a number in K.*

This is true for *cyclic* fields K/κ, but in general not for Abelian fields.

Let us again denote by $\mathrm{Nm} J_K$ the subgroup in J_κ of the idèles which are equivalent to norms. Then the norm symbol $\phi_K(\alpha) = (\alpha, K)$ determines a homomorphic mapping of the factor group $J_\kappa/\mathrm{Nm} J_K$ into the Galois group of K/κ. One would expect that this mapping is one-to-one:

IV. *By means of the norm symbol the factor group $J_\kappa/\mathrm{Nm} J_K$ is isomorphically mapped onto the Galois group of K/κ.*

I, II, III$_c$ (the subscript c indicating restriction to cyclical fields) and IV are the main propositions of what one might call the *norm theory of relative Abelian fields*. They refer to a *given* field K/κ.

There is a second part of the theory, the *class field theory proper*, which is concerned with the manner in which all possible relative Abelian fields K over κ are reflected in the structure of the group J_κ of idèles in K. Each such field K determines, as we have seen, a subgroup NmJ_K of J_κ of finite index. The question arises *which* subgroups J_κ^* of J_κ are generated in this way by Abelian fields K/κ. Clearly the following conditions are necessary:

(1) Every principal idèle is in J_κ^*.

(2) There is a natural number n such that every nth power of an idèle is in J_κ^*.

(3) There is a finite set S of prime spots such that α is in J_κ^* provided α is a unit at every prime spot and equals 1 at the prime spots of S.

The main theorem concerning class fields states that these conditions are also sufficient.

V. *Given a subgroup J_κ^* of J_κ fulfilling the above three conditions* (and therefore, as one readily verifies, of finite index), *there exists a uniquely determined Abelian field K/κ such that* $J_\kappa^* = \mathrm{Nm}J_K$.

We divide the idèles of κ into *classes* by throwing two idèles into the same class if their quotient is in J_κ^*. Then J_κ/J_κ^* is the class group and K is called the corresponding *class field*. The most important example results if one lets J_κ^* consist of the *unit idèles* α whose values $\alpha_\mathfrak{p}$ are $\mathfrak{p}$-adic units at every prime spot $\mathfrak{p}$.[15] Then the classes may be described as the familiar classes of *ideals*: two ideals are put in the same class if their quotient is a principal ideal (α) springing from a number α positive at all real infinite prime spots. The corresponding class field K, the so-called absolute class field, is of relative discriminant 1, and the largest unramified Abelian field over κ (Theorem II). Its degree n over κ is the class number of ideals, its Galois group isomorphic to the class group of ideals in κ (Theorem IV). f being the least power of $\mathfrak{p}$ which lies in the principal class, $\mathfrak{p}$ decomposes into n/f distinct prime ideals in K, each of relative degree f. This last statement does nothing but repeat the norm definition of the class field. Hence the way in which $\mathfrak{p}$ factors in K depends only on the class to which $\mathfrak{p}$ belongs. The easiest way of extending the theory from the case with no ramified prime ideals, which was preponderant in Hilbert's thought, to Takagi's ramified case is by substituting idèles for ideals. Hilbert also stated that every ideal in κ becomes a principal ideal in the absolute class field. It is today possible to show that this is so, by arguments, however, which are far from being fully understood, because this question transcends the domain of Abelian fields.

As was stated above, Hilbert did not prove these theorems in their full generality, but taking his departure from Gauss's theory of genera in quadratic fields and Kummer's investigations he worked his way gradually up from the simplest cases, developing as he went along the necessary machinery of new concepts and propositions about them until he could survey the whole landscape of class fields. We cannot attempt here to give an idea of the highly involved proofs. The completion of the work he left to his successors. The day is probably still far off when we shall have a theory of relative *Galois* number fields of comparable completeness.

Kronecker had shown, and Hilbert found a simpler proof for the fact, that Abelian fields over the rational ground field k are necessarily subfields of the cyclotomic fields, and are thus obtained from the transcendental function $e^{2\pi i x}$ by substituting rational values for the argument x. For Abelian fields over an imaginary quadratic field the so-called complex multiplication of the elliptic and modular functions plays a similar role ("Kronecker's *Jugendtraum*").[16] While Heinrich Weber following in Kronecker's footsteps, and R. Fuëter under Hilbert's guidance, made this dream come true, Hilbert himself began to play with modular functions of several variables which are defined by means of algebraic number fields, and to study their arithmetical implications. He never published these investigations, but O. Blumenthal, and later E. Hecke, used his notes and developed his ideas. The results are provocative, but still far from complete. It is indicative of the fertility of Hilbert's mind at this most productive period of his life that he handed over to his pupils a complex of problems of such fascination as that of the relation between number theory and modular functions.[17]

There remain to be mentioned a particularly simple proof of the transcendence of e and π with which Hilbert opened the series of his arithmetical papers, and the 1909 paper settling Waring's century-old conjecture. I should classify the latter paper among his most original ones, but we can forego considering it more closely because a decade later Hardy and Littlewood found a different approach which yields asymptotic formulas for the number of representations, and it is the Hardy-Littlewood "circle method" which has given rise in recent times to a considerable literature on this and related subjects.[18]

Axiomatics

There could not have been a more complete break than the one dividing Hilbert's last paper on the theory of number fields from his classical book, *Grundlagen der Geometrie*, published in 1899 [*Foundations of Geometry*, Hilbert 1971]. Its only forerunner is a note of the year 1895 on the straight line as the shortest way. But O. Blumenthal records that as early as 1891 Hilbert, discussing a paper on the role of Desargues's and Pappus's theorems read by H. Wiener at a mathematical meeting, made a remark which contains the axiomatic standpoint in a nutshell: "It must be possible to replace in all geometric statements the words *point, line, plane*, by *table, chair, mug*."

The Greeks had conceived of geometry as a deductive science which proceeds by purely logical processes once the few axioms have been established. Both Euclid and Hilbert carry out this program. However, Euclid's list of axioms was still far from being complete; Hilbert's list is complete and there are no gaps in the deductions. Euclid tried to give a descriptive definition of the basic spatial objects and relations with which the axioms deal; Hilbert abstains from such an attempt. All that we must know about those basic concepts is contained in the axioms. The axioms are, as it were, their implicit (necessarily incomplete) definitions. Euclid believed the axioms to be evident; his concern is the real space of the physical universe. But in the deductive system of geometry the evidence, even the truth of the axioms, is irrelevant; they figure rather as hypotheses of which one sets out to develop the logical consequences. Indeed there are many different material interpretations of the basic concepts for which the axioms become true. For instance, the axioms of n-dimensional Euclidean vector geometry hold if a distribution of direct current in a given electric circuit, the n branches of which connect in certain branching points, is called a vector, and Joule's heat produced per unit time by the current is considered the square of the vector's length. In building up geometry on the foundation of its axioms one will attempt to economize as much as possible and thus illuminate the role of the several groups of axioms. Arranged in their natural hierarchy they are the axioms of incidence, order, congruence, parallelism, and continuity. For instance, if the theory of geometric proportions or of the areas of polygons can be established without resorting to the axioms of continuity, this ought to be done.

In all this, though the execution shows the hand of a master, Hilbert is not unique. An outstanding figure among his predecessors is M. Pasch, who had indeed traveled a long way from Euclid when he brought to light the hidden axioms of order and with methodical clarity carried out the deductive program for projective geometry (1882). Others in Germany (I. Schur) and a flourishing school of Italian geometers (Peano, Veronese) had taken up the discussion. With respect to the economy of concepts, Hilbert is more conservative than the Italians: quite deliberately he clings to the Euclidean tradition with its three classes of undefined elements, points, lines, planes, and its relations of incidence, order and congruence of segments and angles. This gives his book a peculiar charm, as if one looked into a face thoroughly familiar and yet sublimely transfigured.

It is one thing to build up geometry on sure foundations, another to inquire into the logical structure of the edifice thus erected. If I am not mistaken, Hilbert is the first who moves freely on this higher "metageometric" level: systematically he studies the mutual independence of his axioms and settles the question of independence from certain limited groups of axioms for some of the most fundamental geometric theorems. His method is the *construction of models*: the model is shown to disagree with one and to satisfy all other axioms; hence the one cannot be a consequence of the others. One outstanding example of this method had been known for a considerable time, the Cayley-Klein model of non-Euclidean geometry. For Veronese's non-Archimedean geometry Levi-Civita (shortly before Hilbert) had constructed a satisfactory arithmetical model. The question of *consistency* is closely related to that of independence. The general ideas appear to us almost banal today, so thoroughgoing has been their influence upon our mathematical thinking. Hilbert stated them in clear and unmistakable language, and embodied them in a work that is like a crystal: an unbreakable whole with many facets. Its artistic qualities have undoubtedly contributed to its success as a masterpiece of science.

In the construction of his models Hilbert displays an amazing wealth of invention. The most interesting examples seem to me the one by which he shows that Desargues's theorem does not follow from the plane incidence axioms, but that the plane incidence axioms combined with Desargues's theorem enable one to embed the plane in a higher dimensional space in which all incidence axioms hold; and then the other example by which he decides whether the Archimedean axiom of continuity is necessary to restore the full congruence axioms after having curtailed them by the exclusion of reflections.[19]

What is the building material for the models? Klein's model of non-Euclidean geometry could be interpreted as showing that he who accepts Euclidean geometry with its points and lines, and so on, can by mere change of nomenclature also get the non-Euclidean geometry.[20] Klein himself preferred another interpretation in terms of projective space. However, Descartes's analytic geometry had long provided a more general and satisfactory answer, of which Riemann, Klein and many others must have been aware: All that we need for our construction is the field of real *numbers*. Hence any contradiction in Euclidean geometry must show up as a contradiction in the arithmetical axioms on which our operations with real numbers are based. Nobody had said that quite clearly before Hilbert. He formulates a complete and simple set of axioms for real numbers. The system of arithmetical axioms will have its exchangeable parts just as the geometric system has. From a purely algebraic standpoint the most important axioms are those characterizing a (commutative or noncommutative) *field*. Any such abstract number field may serve as a basis for the construction of corresponding geometries. Vice versa, one may introduce numbers and their operations in terms of a space satisfying certain axioms; Hilbert's Desarguesian *Streckenrechnung* [distance calculation] is a fine example. In general this reverse process is the more difficult one. The Chicago school under E. H. Moore took up Hilbert's investigations, and in particular O. Veblen did much to reveal the perfect correspondence between the projective spaces obeying a set of simple incidence axioms (and no axioms of order), and the abstractly defined number fields.[21]

What the question of independence literally asks is to make sure that a certain proposition cannot be deduced from other propositions. It seems to require that we make the propositions, rather than the things of which they speak, the object of our investigation, and that as a preliminary we fully analyze the logical mechanism of deduction. The method of models is a wonderful trick to avoid that sort of logical investigations. It pays, however, a heavy price for thus shirking the fundamental issue: it merely reduces everything to the question of consistency for the arithmetical axioms, which is left unanswered. In the same manner *completeness*, which literally means that every general proposition about the objects with which the axioms deal can be decided by inference from the axioms, is replaced by *categoricity* (Veblen), which asserts that any imaginable model is isomorphic to the one model by which consistency is established. In this sense Hilbert proves that there is but "one," the Cartesian geometry, which fulfills all his axioms. Only in the case of G. Fano's and O. Veblen's finite projective spaces, for example, of the projective plane consisting of seven points, the model is a purely combinatorial scheme, and the questions of consistency, independence and completeness can be answered in the absolute sense. Hilbert never seems to have thought of illustrating his conception of the axiomatic method by purely combinatorial schemes, and yet they provide by far the simplest examples.

An approach to the foundations of geometry entirely different from the one followed in his book is pursued by Hilbert in a paper which is one of the earliest documents of set-theoretic topology. From the standpoint of mechanics the central task which geometry ought to perform is that of describing the mobility of a solid. This was the viewpoint of Helmholtz, who succeeded in characterizing the group of motions in Euclidean space by a few simple axioms.[22] The question had been taken up by Sophus Lie in the light of his general theory of continuous groups. Lie's theory depends on certain assumptions of differentiability; to get rid of them is one of Hilbert's Paris Problems. In the paper just mentioned he *does* get rid of them as far as Helmholtz's problem in the plane is concerned. The proof is difficult and laborious; naturally continuity is now the foundation — and not the keystone of the building as it had been in his *Grundlagen* book. Other authors, R. L. Moore, N. J. Lennes, W. Süss, B. v. Kerékjártó, carried the problem further along these topological lines. A half-personal reminiscence may be of interest. Hilbert defines a two-dimensional manifold by means of neighborhoods, and requires that a class of "admissible" one-to-one mappings of a neighborhood upon Jordan domains in an x, y-plane be designated, any two of which are connected by continuous transformations. When I gave a course on Riemann surfaces at Göttingen in 1912, I consulted Hilbert's paper and noticed that the neighborhoods themselves could be used to characterize that class. The ensuing definition was given its final touch by F. Hausdorff; the Hausdorff axioms have become a byword in topology.[23] (However, when it comes to explaining what a *differentiable* manifold is, we are to this day bound to Hilbert's roundabout way; cf. Veblen and Whitehead, *The Foundations of Differential Geometry* [1932]).

The fundamental issue of an *absolute proof of consistency* for the axioms which should include the whole of mathematical analysis, nay even Cantor's set theory in its wildest generality, remained in Hilbert's mind, as a paper read before the International Congress at Heidelberg in 1904 testifies. It shows him on the way, but still far from the goal. Then came the time when integral equations and later physics became his all-absorbing interest. One hears a loud rumbling of the old problem in his Zürich address, *Axiomatisches Denken* ["Axiomatic Thinking"], of 1917. Meanwhile the difficulties concerning the foundations of mathematics had reached a critical stage, and the situation cried for repair. Under the impact of undeniable antinomies in set theory, Dedekind and Frege had revoked their own work on the nature of numbers and arithmetical propositions, Bertrand Russell had pointed out the hierarchy of types which, unless one decides to "reduce" them by sheer force, undermine the arithmetical theory of the continuum; and finally L. E. J. Brouwer by his intuitionism had opened our eyes and made us see how far generally accepted mathematics goes beyond such statements as can claim real meaning and truth founded on evidence. I regret that in his opposition to Brouwer, Hilbert never openly acknowledged the profound debt which he, as well as all other mathematicians, owe Brouwer for this revelation.

Hilbert was not willing to make the heavy sacrifices which Brouwer's standpoint demanded, and he saw, at least in outline, a way by which the cruel mutilation could be avoided. At the same time he was alarmed by signs of wavering loyalty within the ranks of mathematicians some of whom openly sided with Brouwer. My own article on the *Grundlagenkrise* [crisis of the foundations], written in the excitement of the first postwar years in Europe, is indicative of the mood.[24] Thus Hilbert returns to the problem of foundations in earnest. He is convinced that complete certainty can be restored without "committing treason to our science." There is anger and determination in his voice when he proposes "*die Grundlagenfragen einfürallemal aus der Welt zu schaffen* [to dispose of the foundational questions once and for all]." "Forbidding a mathematician to make use of the principle of excluded middle," says he, "is like forbidding an astronomer his telescope or a boxer the use of his fists."

Hilbert realized that the mathematical statements themselves could not be made the subject of a mathematical investigation whose aim is to answer the question of their consistency in its primitive sense, lest they be first reduced to mere *formulas*. Algebraic formulas like $a + b = b + a$ are the most familiar examples. The process of deduction by which formulas previously obtained give rise to new formulas must be described without reference to any meaning of the formulas. The deduction starts from certain initial formulas, the axioms, which must be written out explicitly. Whereas in his *Grundlagen der Geometrie* the meaning of the geometric terms had become irrelevant, but the meaning of logical terms, as "and," "not," "if then," had still to be understood, now every trace of meaning is obliterated. As a consequence, logical symbols like $\rightarrow$ in $a \rightarrow b$, read: a implies b, enter into the formulas. Hilbert fully agrees with Brouwer in that the great majority of mathematical propositions are not "real" ones conveying a definite meaning verifiable in the light of evidence. But he insists that the non-real, the "ideal propositions" are indispensable in order to give our mathematical system "completeness." Thus he parries Brouwer, who had asked us to give up what is meaningless, by relinquishing the pretension of meaning altogether, and what he tries to establish is not *truth* of the individual mathematical proposition, but *consistency* of the system. The game of deduction when played according to rules, he maintains, will never lead to the formula $0 \neq 0$. In this sense, and in this sense only, he promises to salvage our cherished classical mathematics in its entirety.

For those who accuse him of degrading mathematics to a mere game he points first to the introduction of ideal elements for the sake of completeness as a common method in all mathematics — for example, of the ideal points outside an accessible portion of space, without which space would be incomplete—; secondly, to the neighboring science of physics where likewise not the individual statement is verifiable by experiment, but in principle only the system as a whole can be confronted with experience.

But how to make sure that the "game of deduction" never leads to a contradiction? Shall we prove this by the same mathematical method the validity of which stands in question, namely by deduction from axioms? This would clearly involve a regress *ad infinitum*. It must have been hard on Hilbert, the axiomatist, to acknowledge that the insight of consistency is rather to be attained by *intuitive reasoning* which is based on evidence and not on axioms. But after all, it is not surprising that ultimately the mind's seeing eye must come in. Already in communicating the rules of the game we must count on understanding. The game is played in silence, but the rules must be *told* and any reasoning about it, in particular about its consistency, communicated by *words*. Incidentally, in describing the indispensable intuitive basis for his *Beweistheorie* [theory of proof] Hilbert shows himself an accomplished master of that, alas, so ambiguous medium of communication, language. With regard to what he accepts as evident in this "metamathematicial" reasoning, Hilbert is more papal than the pope, more exacting than either Kronecker or Brouwer. But it cannot be helped that our reasoning in following a hypothetic sequence of formulas leading up to the formula $0 \neq 0$ is carried on in hypothetic generality and uses that type of evidence which a formalist would be tempted to brand as application of the principle of complete induction. Elementary arithmetics can be founded on such intuitive reasoning as Hilbert himself describes, but we need the formal apparatus of variables and "quantifiers" to invest the infinite with the all important part that it plays in higher mathematics. Hence Hilbert prefers to make a clear cut: he becomes strict formalist in mathematics, strict intuitionist in metamathematics.

It is perhaps possible to indicate briefly how Hilbert's formalism restores the *principle of the excluded middle* which was the main target of Brouwer's criticism. Consider the infinite sequence of numbers $0, 1, 2, \cdots$. Any property A of numbers (for example, "being prime") may be represented by a propositional function $A(x)$ ("x is prime"), from which a definite proposition $A(b)$ arises by substituting a concrete number b for the variable x ("6 is prime").[25] Accepting the principle which Brouwer denies and to which Hilbert wishes to hold on, that (i) either there exists a number x for which $A(x)$ holds, or (ii) $A(x)$ holds for no x, we can find a "representative" r for the property A, a number such that $A(b)$ implies $A(r)$ whatever the number b, $A(b) \to A(r)$. Indeed, in the alternative (i) we choose r as one of the numbers x for which $A(x)$ holds, in the alternative (ii) at random. Thus Aristides is the representative of honesty; for, as the Athenians said, if there is any honest man it is Aristides. Assuming that we know the representative we can decide the question whether there is an honest man or whether all are dishonest by merely looking at *him*: if he is dishonest everybody is. In the realm of numbers we may even make the choice of the representative unique, in case (i) choosing $x = r$ as the *least* number for which $A(x)$ holds, and $r = 0$ in the opposite case (ii). Then r arises from A by a certain operator $\rho_x, r = \rho_x A(x)$, applicable to every imaginable property A. A propositional function may contain other variables $y, z, \cdots$ besides x. Therefore it is necessary to attach an index x to ρ, just as in integrating one must indicate with respect to which variable one integrates. ρ_x eliminates the variable x; for instance $\rho_x A(x, y)$ is a function of y alone. The word quantifier is in use for this sort of operator. Hence we write our axiom as follows:

(8) $$A(b) \to A(\rho_x A(x)).$$

It is immaterial whether we fix the representative in the unique manner described above; our specific rule would not fit anyhow unless x ranges over the numbers $0, 1, 2, \cdots$. Instead we imagine a quantifier ρ_x of universal applicability which, as it were, selects the representative for us. Zermelo's axiom of choice is thus woven into the principle of the excluded middle.[26] It is a bold step; but the bolder the better, as long as it can be shown that we keep within the bounds of consistency!

In the formalism, propositional functions are replaced by *formulas* the handling of which must be described without reference to their meaning. In general, variables $x, y, \cdots$ will occur among the symbols of a formula $\mathfrak{A}$. We say that the symbol ρ_x *binds* the variable x in the formula $\mathfrak{A}$ which follows the symbol[27] and that x occurs *free* in a formula wherever it is not bound by a quantifier with index x. $x, y, \to, \rho_x$ are symbols entering into the formulas; the German letters are no such symbols, but are used for communication. It is more natural to describe our critical axiom (8) as a rule for the formation of axioms. It says: take any formula $\mathfrak{A}$ in which only the variable x occurs free, and any formula b without free variables, and by means of them build the formula

(9) $$\mathfrak{A}(b) \to \mathfrak{A}(\rho_x \mathfrak{A}).$$

Here $\mathfrak{A}(b)$ stands for the formula derived from $\mathfrak{A}$ by putting in the entire formula b for the variable x *wherever x occurs free*.

In this way formulas may be *obtained* as *axioms* according to certain rules. *Deduction* proceeds by the rule of syllogism: From two formulas a and $a \to b$ *previously obtained*, in the second of which the first formula reappears at the left of the symbol $\to$, one *obtains* the formula b.

How does Hilbert propose to show that the game of deduction will never lead to the formula $0 \neq 0$? Here is the basic idea of his procedure. As long as one deals with "finite" formulas only, formulas from which the quantifiers $\rho_x, \rho_y, \cdots$ are absent, one can decide whether they are true or false by merely looking at them. With the entrance of ρ such a descriptive valuation of formulas becomes impossible: evidence ceases to work. But a concretely given deduction is a sequence of formulas in which only a limited number of instances of the axiomatic rule (9) will turn up. Let us assume that the only quantifier which occurs is ρ_x and wherever it occurs it is followed by the *same finite* formula $\mathfrak{A}$, so that the instances of (9) are of the form

$$(10) \qquad \mathfrak{A}(\mathfrak{b}_1) \to \mathfrak{A}(\rho_x \mathfrak{A}), \cdots, \mathfrak{A}(\mathfrak{b}_h) \to \mathfrak{A}(\rho_x \mathfrak{A}).$$

Assume, moreover, $\mathfrak{b}_1, \cdots, \mathfrak{b}_h$ to be finite. We then carry out a *reduction*, replacing $\rho_x \mathfrak{A}$ by a certain finite $\mathfrak{r}$ wherever it occurs as part of a formula in our sequence. In particular, the formulas (10) will change into

$$(11) \qquad \mathfrak{A}(\mathfrak{b}_1) \to \mathfrak{A}(\mathfrak{r}), \cdots, \mathfrak{A}(\mathfrak{b}_h) \to \mathfrak{A}(\mathfrak{r}).$$

We now see how to choose $\mathfrak{r}$. If by examining the finite formulas $\mathfrak{A}(\mathfrak{b}_1), \cdots, \mathfrak{A}(\mathfrak{b}_h)$ one after the other, we find one that is true, say $\mathfrak{A}(\mathfrak{b}_3)$, then we take $\mathfrak{b}_3$ for $\mathfrak{r}$. If every one of them turns out to be false, we choose $\mathfrak{r}$ at random. Then the h reduced formulas (11) are "true" and our hypothesis that the deduction leads to the false formula $0 \neq 0$ is carried *ad absurdum*. The salient point is that a concretely given deduction makes use of a limited number of explicitly exhibited individuals $\mathfrak{b}_1, \cdots, \mathfrak{b}_h$ only. If we make a wrong choice, for example, by choosing Alcibiades rather than Aristides as the representative of incorruptibility, our mistake will do no harm as long as the few people (out of the infinite Athenian crowd) with whom we actually deal are all corruptible.

A slightly more complicated case arises when we permit the $\mathfrak{b}_1, \cdots, \mathfrak{b}_h$ to contain ρ_x, but always followed by the same $\mathfrak{A}$. Then we first make a *tentative* reduction replacing $\rho_x \mathfrak{A}$ by the number 0, say. The formulas $\mathfrak{b}_1, \cdots, \mathfrak{b}_h$ are thus changed into reduced finite formulas $\mathfrak{b}_1^0, \cdots, \mathfrak{b}_h^0$ and (10) into

$$\mathfrak{A}(\mathfrak{b}_1^0) \to \mathfrak{A}(0), \cdots, \mathfrak{A}(\mathfrak{b}_h^0) \to \mathfrak{A}(0).$$

This reduction will do unless $\mathfrak{A}(0)$ is false and at the same time one of the $\mathfrak{A}(\mathfrak{b}_1^0), \cdots, \mathfrak{A}(\mathfrak{b}_h^0)$, say $\mathfrak{A}(\mathfrak{b}_3^0)$, is true. But then we have in $\mathfrak{b}_3^0$ a perfectly legitimate representative of $\mathfrak{A}$, and a second reduction which replaces $\rho_x \mathfrak{A}$ by $\mathfrak{b}_3^0$ will work out all right.

However, this is only a modest beginning of the complications awaiting us. Quantifiers $\rho_x, \rho_y, \cdots$ with different variables and applied to different formulas will be piled one upon the other. We make a tentative reduction; it will go wrong in certain places and from that failure we learn how to correct it. But the corrected reduction will probably go wrong at other places. We seem to be driven around in a vicious circle, and the problem is to direct our consecutive corrections in such a manner as to obtain assurance that finally a reduction will result that makes good at all places in our given sequence of formulas. Nothing has contributed more to revealing the circle-like character of the usual transfinite arguments of mathematics than these attempts to make sure of consistency in spite of all circles.

The symbolism for the formalization of mathematics as well as the general layout and first steps of the proof of consistency are due to Hilbert himself. The program was further advanced by younger collaborators, P. Bernays, W. Ackermann, J. von Neumann. The last two proved the consistency of "arithmetics," of that part in which the dangerous axiom about the conversion of predicates into sets is not yet admitted. A gap remained which seemed harmless at the time, but already detailed plans were drawn up for the invasion of analysis. Then came a catastrophe: assuming that consistency is established, K. Gödel showed how to construct arithmetical propositions which are evidently true and yet not deducible within the formalism. His method applies to Hilbert's as well as any other not too limited formalism. Of the two fields, the field of formulas obtainable in Hilbert's formalism and the field of real propositions that are evidently true, neither contains the other (provided consistency of the formalism can be made evident). Obviously *completeness* of a formalism in the absolute sense in which Hilbert had envisaged it was now out of the question. When G. Gentzen later closed the gap in the consistency proof for arithmetics, which Gödel's discovery had revealed to be serious indeed, he succeeded in doing so only by substantially lowering Hilbert's standard of evidence.[28] The boundary line of what is intuitively trustworthy once more became vague. As all hands were needed to defend the homeland of arithmetics, the invasion of analysis never came off, to say nothing of general set theory.

This is where the problem now stands; no final solution is in sight. But whatever the future may bring, there is no doubt that Brouwer and Hilbert raised the problem of the foundations of mathematics to a new level. A return to the standpoint of Russell-Whitehead's *Principia Mathematica* is unthinkable.

Hilbert is the champion of axiomatics. The axiomatic attitude seemed to him one of universal significance, not only for mathematics, but for all sciences. His investigations in the field of physics are conceived in the axiomatic spirit. In his lectures he liked to illustrate the method by examples taken from biology, economics, and so on. The modern epistemological interpretation of science has been profoundly influenced by him. Sometimes when he praised the axiomatic method he seemed to imply that it was destined to obliterate completely the constructive or genetic method. I am certain that, at least in later life, this was not his true opinion. For whereas he deals with the primary mathematical objects by means of the axioms of his symbolic system, the formulas are constructed in the most explicit and finite manner. In recent times the axiomatic method has spread from the roots to all branches of the mathematical tree. Algebra, for one, is permeated from top to bottom by the axiomatic spirit. One may describe the role of axioms here as the subservient one of fixing the range of variables entering into the explicit constructions. But it would not be too difficult to retouch the picture so as to make the axioms appear as the masters. An impartial attitude will do justice to both sides; not a little of the attractiveness of modern mathematical research is due to a happy blending of axiomatic and genetic procedures.

Integral equations

Between the two periods during which Hilbert's efforts were concentrated on the foundations, first of geometry, then of mathematics in general, there lie twenty long years devoted to analysis and physics.

In the winter of 1900–1901 the Swedish mathematician E. Holmgren reported in Hilbert's seminar on Fredholm's first publications on integral equations, and it seems that Hilbert caught fire at once. The subject has a long and tortuous history, beginning with Daniel Bernoulli. The mathematicians' efforts to solve the (mechanical, acoustical, optical, electromagnetical) problem of the oscillations of a continuum and the related boundary value problems of potential theory span a period of two centuries. Fourier's *Théorie analytique de la chaleur* (1822) [*Analytical Theory of Heat*] is a landmark. H. A. Schwarz proved for the first time (1885) the existence of a proper oscillation in two and more dimensions by constructing the fundamental frequency of a membrane. The last decade of the nineteenth century saw Poincaré on his way to the development of powerful function-theoretic methods; C. Neumann and he came to grips with the harmonic boundary problem; Volterra studied that type of integral equations which now bears his name, and for linear equations with infinitely many unknowns Helge von Koch developed the infinite determinants. Most scientific discoveries are made when "their time is fulfilled"; sometimes, but seldom, a genius lifts the veil decades earlier than could have been expected. Fredholm's discovery has always seemed to me one that was long overdue when it came. What could be more natural than the idea that a set of linear equations connected with a discrete set of mass points gives way to an integral equation when one passes to the limit of a continuum? But the fact that in the simpler cases a differential rather than an integral equation results in the limit riveted the mathematicians' attention for two hundred years on differential equations!

It must be said, however, that the simplicity of Fredholm's results is due to the particular form of his equation, on which it was hard to hit without the guidance of the problems of mathematical physics to which he applied it:

$$x(s) - \int_0^1 K(s,t)\,x(t)\,dt = f(s) \qquad (0 \leqq s \leqq 1).$$

Indeed the linear operator which in the left member operates on the unknown function x producing a given f, $(E - K)x = f$, consists of two parts, the identity E and the integral operator K, which in a certain sense is weak compared to E. Fredholm proved that for this type of integral equation the two main facts about

n linear equations with the same number n of unknowns hold: (1) The homogeneous equation $[f(s) = 0]$ has a finite number of linearly independent solutions $x(s) = \phi_1(s), \cdots, \phi_h(s)$, and the homogeneous equation with the transposed kernel $K'(s,t) = K(t,s)$ has the same number of solutions, $\psi_1(s), \cdots, \psi_h(s)$. (2) The nonhomogeneous equation is solvable if and only if the given f satisfies the h linear conditions

$$\int_0^1 f(s)\,\psi_i(s)\,ds = 0 \quad (i = 1, \cdots, h).$$

Following an artifice used by Poincaré, Fredholm introduces a parameter λ replacing K by λK and obtains a solution in the form familiar from finite linear equations, namely as a quotient of two determinants of H. von Koch's type, either of which is an entire function of the parameter λ.

Hilbert saw two things: (1) after having constructed Green's function K for a given region G and for the potential equation $\Delta u = 0$ by means of a Fredholm equation on the boundary, the differential equation of the oscillating membrane $\Delta \phi + \lambda \phi = 0$ changes into a homogeneous integral equation

$$\phi(s) - \lambda \int K(s,t)\,\phi(t)\,dt = 0$$

with the symmetric kernel $K, K(t,s) = K(s,t)$ (in which the parameter λ is no longer artificial but of the very essence of the problem); (2) the problem of ascertaining the "eigenvalues" λ and "eigenfunctions" $\phi(s)$ of this integral equation is the analogue for integrals of the transformation of a quadratic form of n variables onto principal axes. Hence the corresponding theorem for the quadratic integral form

$$(12) \qquad \int_0^1 \int_0^1 K(s,t)\,x(s)x(t)\,ds\,dt$$

with an *arbitrary symmetric kernel* K must provide the general foundation for the theory of oscillations of a continuous medium. If others saw the same, Hilbert saw it at least that much more clearly that he bent all his energy on proving that proposition, and he succeeded by the same direct method which about 1730 Bernoulli had applied to the oscillations of a string: passage to the limit from the algebraic problem. In carrying out the limiting process he had to make use of the Koch-Fredholm determinant. He finds that there is a sequence of eigenvalues $\lambda_1, \lambda_2, \cdots$ tending to infinity, $\lambda_n \to \infty$ for $n \to \infty$, and an orthonormal set of corresponding eigenfunctions $\phi_n(s)$

$$\phi_n(s) - \lambda_n \int K(s,t)\,\phi_n(t)\,dt = 0,$$

$$\int_0^1 \phi_m(s)\,\phi_n(s)\,ds = \delta_{mn},$$

such that

$$\int_0^1 \int_0^1 K(s,t)\,x(s)x(t)\,ds\,dt = \sum \xi^2/\lambda_n,$$

ξ_n being the Fourier coefficient $\int_0^1 x(s)\phi_n(s)\,ds$. The theory implies that every function of the form

$$y(s) = \int_0^1 K(s,t)\,x(t)\,dt$$

may be expanded into a uniformly convergent Fourier series in terms of the eigenfunctions ϕ_n,

$$y(s) = \sum \eta_n\,\phi_n(s), \qquad \eta_n = \int_0^1 y(s)\,\phi_n(s)\,ds.$$

Hilbert's passage to the limit is laborious. Soon afterwards E. Schmidt in a Göttingen thesis found a simpler and more constructive proof for these results by adapting H. A. Schwarz's method invented twenty years before to the needs of integral equations.

From finite forms the road leads either to integrals or to infinite series. Therefore Hilbert considered the same problem of orthogonal transformation of a given quadratic form

(13) $$\sum K_{mn}x_m x_n$$

into a form of the special type

(14) $$\kappa_1 \xi_1^2 + \kappa_2 \xi_2^2 + \cdots \qquad (\kappa_n = 1/\lambda_n \to 0)$$

also for infinitely many (real) variables $(x_1, x_2, \cdots)$ or vectors x in a space of a denumerable infinity of dimensions. Only such vectors are admitted as have a finite length $|x|$,

$$|x|^2 = x_1^2 + x_2^2 + \cdots;$$

they constitute what we now call the Hilbert space. The advantage of Hilbert space over the "space" of all continuous functions $x(s)$ lies in a certain property of completeness, and due to this property one can establish "complete continuity" as the necessary and sufficient condition for the transformability of a given quadratic form K, (13), into (14), by following an argument well known in the algebraic case: one determines $\kappa_1, \kappa_2, \cdots$ as the consecutive maxima of K on the "sphere" $|x|^2 = 1$.

As suggested by the theorem concerning a quadratic integral form, the link between the space of functions $x(s)$ and the Hilbert space of vectors $(x_1, x_2, \cdots)$ is provided by an arbitrary *complete* orthonormal system $u_1(s), u_2(s), \cdots$ and expressed by the equations

$$x_n = \int_0^1 x(s)\,u_n(s)\,ds.$$

Bessel's inequality states that the square sum of the Fourier coefficients x_n is less than or equal to the square integral of $x(s)$. The relation of completeness, first introduced by A. Hurwitz and studied in detail by W. Stekloff, requires that in this inequality the *equality* sign prevail. Thus the theorem on quadratic forms of infinitely many variables at once gives the corresponding results about the eigenvalues and eigenfunctions of symmetric kernels $K(s,t)$ — or would do so if one could count on the uniform convergence of $\sum x_n u_n(s)$ for any given vector $(x_1, x_2, \cdots)$ in Hilbert space. In the special case of an eigenvector of that quadratic form (13) which corresponds to the integral form (12),

$$x_n = \lambda \sum_m K_{nm} x_m.$$

Hilbert settles this point by forming the uniformly convergent series

$$\lambda \sum_m x_m \int_0^1 K(s,t) \, u_m(t) \, dt$$

which indeed yields a continuous function $\phi(s)$ with the nth Fourier coefficient

$$\lambda \sum K_{nm} x_m = x_n,$$

and thus obtains the eigenfunction of $K(s,t)$ for the eigenvalue λ. Soon afterwards, under the stimulus of Hilbert's investigations, E. Fischer and F. Riesz proved their well-known theorem that the space of all functions $x(s)$ the square of which has a finite Lebesgue integral enjoys the same property of completeness as Hilbert space, and hence one is mapped isomorphically upon the other in a one-to-one fashion by means of a complete orthonormal system $u_n(s)$. I mention these details because the historic order of events may have fallen into oblivion with many of our younger mathematicians, for whom Hilbert space has assumed that abstract connotation which no longer distinguishes between the two realizations, the square integrable functions $x(s)$ and the square summable sequences $(x_1, x_2, \cdots)$. I think Hilbert was wise to keep within the bounds of continuous functions when there was no actual need for introducing Lebesgue's general concepts.

Perhaps Hilbert's greatest accomplishment in the field of integral equations is his extension of the theory of spectral decomposition from the completely continuous to the so-called *bounded* quadratic forms. He finds that then the point spectrum will in general have condensation points and a continuous spectrum will appear beside the point spectrum. Again he proceeds by directly carrying out the transition to the limit, letting the number of variables $x_1, x_2, \cdots$ increase ad infinitum. Again, not long afterwards, simpler proofs for his results were found.

While thus advancing the boundaries of the general theory, he did not lose sight of the ordinary and partial differential equations from which it had sprung. Simultaneously with the young Italian mathematician Eugenio Elia Levi he developed the parametrix method as a bridge between differential and integral equations. For a given elliptic differential operator Δ^* of the second order, the parametrix $K(s,t)$ is a sort of qualitative approximation of Green's function, depending like the latter on an argument point s and a parameter point t. It is supposed to possess the right kind of singularity for $s = t$ so that the nonhomogeneous equation $\Delta^* u = f$ for

$$u = K\rho, \qquad u(s) = \int K(s,t)\,\rho(t)\,dt$$

gives rise to the integral equation $\rho + L\rho = f$ for the density ρ, with a kernel $L(s,t) = \Delta_s^* K(s,t)$ regular enough at $s = t$ for Fredholm's theory to be applicable. It is important to give up the assumption that K satisfies the equation $\Delta^* K = 0$, because in general a fundamental solution will not be known for the given differential operator Δ^*. In order not to be bothered by boundary conditions, Hilbert assumes the domain of integration to be a *compact* manifold, like the surface of a sphere, and finds that the method works if the parametrix, besides having the right kind of singularity, is symmetric with respect to argument and parameter.

What has been said should be enough to make clear that in the terrain of analysis a rich vein of gold had been struck, comparatively easy to exploit and not soon to be exhausted. The linear equations of infinitely many unknowns had to be investigated further (E. Schmidt, F. Riesz, O. Toeplitz, E. Hellinger, and others); the continuous spectrum and its appearance in integral equations with "singular" kernels awaited closer analysis (E. Hellinger, T. Carleman); ordinary differential equations, with regular or singular boundaries, of second or of higher order, received their due share of attention (A. Kneser, E. Hilb, G. D. Birkhoff, M. Bôcher, J. D. Tamarkin, and many others).[29] It became possible to develop such asymptotic laws for the distribution of eigenvalues as were required by the thermodynamics of radiation (H. Weyl, R. Courant). Expansions in terms of orthogonal functions were studied independently of their origin in differential or integral equations. New light fell upon Stieltjes's continued fractions and the problem of momentum. The most ambitious began to attack nonlinear integral equations. A large international school of young mathematicians gathered around Hilbert and integral equations became the fashion of the day, not only in Germany, but also in France where great masters like E. Picard and Goursat paid their tributes, in Italy and on this side of the Atlantic. Many good papers were written, and many mediocre ones. But the total effect was an appreciable change in the aspect of analysis.

Remarkable are the applications of integral equations outside the field for which they were invented. Among them I mention the following three: (1) Riemann's problem of determining n analytic functions $f_1(z), \cdots, f_n(z)$, regular except at a finite number of points, which by analytic continuation around these points preassigned linear transformations. The problem was solved by Hilbert himself, and subsequently in a simpler and more complete form by J. Piemelj. (A very special case of it is the existence of algebraic functions on a Riemann surface if that surface is given as a covering surface of the complex z-plane.) Investigations by G. D. Birkhoff on matrices of analytic functions lie in the same line. (2) Proof for the completeness of the irreducible representations of a compact continuous group. This is an indispensable tool for the approach to the general theory of invariants by means of Adolf Hurwitz's integration method, and with its refinements and generalizations plays an important role in modern group-theoretic research, including H. Bohr's theory of almost periodic functions.[30] Contact is thus made with Hilbert's old friend, the theory of invariants. (3) Quite recently Hilbert's parametrix method has served to establish the central existence theorem in W. V. D. Hodge's theory of harmonic integrals in compact Riemannian spaces.[31]

The story would be dramatic enough had it ended here. But then a sort of miracle happened: the spectrum theory in Hilbert space was discovered to be the adequate mathematical instrument of the new quantum physics inaugurated by Heisenberg and Schrödinger in 1923. This latter impulse led to a reexamination of the entire complex of problems with refined means (J. von Neumann, A. Wintner, M. H. Stone, K. Friedrichs). As J. von Neumann was Hilbert's collaborator toward the close of that epoch when his interest was divided between quantum physics and foundations, the historic continuity with Hilbert's own scientific activities is unbroken, even for this later phase. What has become of the theory of abstract spaces and their linear operators in our times lies beyond the bounds of this report.

A picture of Hilbert's "analytic" period would be incomplete without mentioning a second motif, *calculus of variations*, which crossed the dominating one of integral equations. The "theorem of independence" with which he concludes his Paris survey of mathematical problems (1900) is an important contribution to the formal apparatus of that calculus. But of much greater consequence was his audacious direct assault on the functional maxima and minima problems. The whole finely wrought machinery of the calculus of variations is here consciously set aside. He proposes instead to construct the minimizing function as the limit of a sequence of functions for which the value of the integral under investigation tends to its minimum value. The classical example is Dirichlet's integral in a two-dimensional region G,

$$D[u] = \iint_G \left\{ \left(\frac{\partial u}{\partial x}\right)^2 + \left(\frac{\partial u}{\partial y}\right)^2 \right\} dx\, dy.$$

Admitted are all functions u with continuous derivatives which have given boundary values, d being the lower limit of $D[u]$ for admissible u, one can ascertain a sequence of admissible functions u_n such that $D[u_n] \to d$ with $n \to \infty$. One cannot expect the u_n themselves to converge; rather they have to be prepared for convergence by the smoothing process of integration. As the limit function will be harmonic and the value of the harmonic function at any point P equals its mean value over any circle K around P, it seems best to replace $u_n(P)$ by its mean value in K, with the expectation that this mean value will converge toward a number $u(P)$ which is independent of the circle and in its dependence on P solves the minimum problem. Besides integration Hilbert uses the process of sifting a suitable subsequence from the u_n before passing to the limit. Owing to the simple inequality

$$\{D[u_m - u_n]\}^{1/2} \leqq \{D[u_m] - d\}^{1/2} + \{D[u_m] - d)\}^{1/2}$$

discovered by S. Zaremba this second step is unnecessary.

Hilbert's method is even better suited for problems in which the boundary does not figure so prominently as in the boundary value problems. By a slight modification one is able to include point singularities, and Hilbert thus solved the fundamental problem for flows on Riemann surfaces, providing thereby the necessary foundation for Riemann's own approach to the theory of Abelian integrals, and he further showed that Poincaré's and Koebe's fundamental theorems on uniformization could be established in the same way. We should be much better off in number theory if methods were known which are as powerful for the construction of relative Abelian and Galois fields over given algebraic number fields as the Riemann-Hilbert transcendental method proves to be for the analogous problems in the fields of algebraic functions! Its wide application in the theory of conformal mapping and of minimal surfaces is revealed by the work of the man who was Hilbert's closest collaborator in the direction of mathematical affairs at Göttingen for many years, Richard Courant.[32] Of a more indirect character, but of considerable vigor, is the influence of Hilbert's ideas upon the whole trend of the modern development of the calculus of variations; in Europe Carathéodory, Lebesgue, Tonelli could be mentioned among others, in this country the chain reaches from O. Bolza's early to M. Morse's most recent work.

Physics

Already before Minkowski's death in 1909, Hilbert had begun a systematic study of theoretical physics, in close collaboration with his friend who had always kept in touch with the neighboring science. Minkowski's work on relativity theory was the first fruit of these joint studies. Hilbert continued them through the years, and between 1910 and 1930 often lectured and conducted seminars on topics of physics. He greatly enjoyed this widening of his horizon and his contact with physicists, whom he could meet on their own ground. The harvest however can hardly be compared with his achievements in pure mathematics. The maze of experimental facts which the physicist has to take into account is too manifold, their expansion too fast, and their aspect and relative weight too changeable for the axiomatic method to find a firm enough foothold, except in the thoroughly consolidated parts of our physical knowledge. Men like Einstein or Niels Bohr grope their way in the dark toward their conceptions of general relativity or atomic structure by another type of experience and imagination than those of the mathematician, although no doubt mathematics is an essential ingredient. Thus Hilbert's vast plans in physics never matured.

But his application of integral equations to kinetic gas theory and to the elementary theory of radiation were notable contributions. In particular, his asymptotic solution of Maxwell-Boltzmann's fundamental equation in kinetic gas theory, which is an integral equation of the second order, clearly separated the two layers of phenomenological physical laws to which the theory leads; it has been carried out in more detail by the physicists and applied to several concrete problems. In his investigations on general relativity Hilbert combined Einstein's theory of gravitation with G. Mie's program of pure field physics. For the development of the theory of general relativity at that stage, Einstein's more sober procedure, which did not couple the theory with Mie's highly speculative program, proved the more fertile. Hilbert's endeavors must be looked upon as a forerunner of a unified field theory of gravitation and electromagnetism. However, there was still much too much arbitrariness involved in Hilbert's Hamiltonian function; subsequent attempts (by Weyl, Eddington, Einstein himself, and others) aimed to reduce it. Hopes in the Hilbert circle ran high at that time; the dream of a universal law accounting both for the structure of the cosmos as a whole, and of all the atomic nuclei, seemed near fulfillment. But the problem of a unified field theory stands to this day as an unsolved problem; it is almost certain that a satisfactory solution will have to include the material waves (the Schrdinger-Dirac ψ for the electron, and similar field quantities for the other nuclear particles) besides gravitation and electromagnetism, and that its mathematical frame will not be a simple enlargement of that of Einstein's now classical theory of gravitation.

Hilbert was not only a great scholar, but also a great teacher. Witnesses are his many pupils and assistants, whom he taught the handicraft of mathematical research by letting them share in his own work and its overflow, and then his lectures, the notes of many of which have found their way from Göttingen into public and private mathematical libraries. They covered an extremely wide range. The book he published with S. Cohn-Vossen on *Anschauliche Geometrie* [*Geometry and the Imagination*, Hilbert and Cohn-Vossen 1999], is an outgrowth of his teaching activities. Going over the impressive list attached to his *Collected Papers* (vol. 3, p. 430) one is struck by the considerable number of courses on general topics like "Knowledge and Thinking," "On the Infinite," "Nature and Mathematics." His speech was fairly fluent, not as hesitant as Minkowski's, and far from monotonous. He had no difficulty in finding the pregnant words, and liked to emphasize short pivotal phrases by repeating them several times. On the whole, his lectures were a faithful reflection of his spirit; direct, intense; how could they fail to be inspiring?

Notes

[Published as Weyl 1944b; *WGA* 4:130–172. Weyl's punctuation has been preserved but compound words like "eigen function" have been rendered as one word, following present-day usage.]

[1] [For these academic ranks, see above, 12, 66.]

[2] How Georg Cantor himself in his excitability suffered from Kronecker's opposition is shown by his violent outbursts in letters to Mittag-Leffler; see A. Schoenflies, *Die Krisis in Cantors mathematischem Schaffen*, Acta Math. vol. 50 (1923), pp. 1–23.

[3] [Hilbert's 1922 essay (*HGA* 3:157–177), translated in Ewald 1996, 2:1115–1134, directly addresses and critiques Weyl's position.]

[4] [Regarding Hilbert's *Zahlbericht*, see above, 91n2.]

[5] [Robert Maynard Hutchins, president of the University of Chicago (1929–1945), sought to promote education based on study of the great books as the center of what Weyl refers to as the "universality of letters" or knowledge, a Latin phrase describing the original purpose of the university.]

[6] [Regarding Hilbert's list of problems, see Gray 2000.]

[7] [For Dirichlet's principle, see below, 180.]

[8] [For the material in this and the following section on algebraic number fields, Kleiner 2007 gives very helpful brief overviews and summaries on a fairly elementary level.]

[9] The example of finite groups is used here as an illustration only. Indeed, a direct elementary proof of the first main theorem for finite groups that makes no use of Hilbert's principle (A) has been given by E. Noether, *Math. Ann.* vol. 77 (1916), p. 89. In dividing by N we have assumed the field k to be of characteristic zero.

[10] B. L. van der Waerden's book *Moderne Algebra*, vol. 2, 2nd ed., 1940 [van der Waerden 1970], gives on pp. 1–72 an excellent account of the general algebraic concepts and facts with which we are here concerned.

[11] I recommend to the reader's attention a brief résumé of his invariant-theoretic work which Hilbert himself wrote for the International Mathematical Congress held at Chicago in conjunction with the World Fair in 1893; *Collected Papers*, vol. 2, item 23.

[12] O. Veblen, *Trans. Amer. Math. Soc.* vol. 7 (1906), pp. 197–199. E. Artin and O. Schreier, *Abh. Math. Sem. Hamburgischen Univ.* vol. 5 (1926), pp. 85–99; E. Artin, ibid. pp. 100–115.

[13] The latest account of the theory is C. Chevalley's paper "La théorie du corps de classes," *Ann. of Math.* vol. 41 (1940), pp. 394–418.

[14] [For a good elementary introduction to p-adic numbers, see Gouvêa 1997.]

[15] At the (real) infinite prime spots the positive numbers are considered the units.

[16] [Kronecker's "youthful dream" was to describe the abelian extensions of an arbitrary number field; among his famous problems, Hilbert listed this as number 12, which remains unsolved; see Gray 2000.]

[17] R. Fuëter, *Singuläre Moduln und complexe Multiplication*, 2 vols., Leipzig, 1924, 1927; cf. also H. Hasse, *J. Reine Angew. Math.* vol. 157 (1927), pp. 115–139. O. Blumenthal, *Math. Ann.* vol. 56 (1903), pp. 509–548, vol. 58 (1904), pp. 497–527. E. Hecke, *Math. Ann.* vol. 71 (1912), pp. 1–37, vol. 74 (1913), pp. 465–510.

[18] It must suffice here to quote the first paper in this line: G. H. Hardy and J. E. Littlewood, *Quart. J. Math.* vol. 48 (1919), pp. 272–293, and its latest successor which carries Waring's theorem over to arbitrary algebraic fields: C. L. Siegel, *Amer. J. Math.* vol. 66 (1944), pp. 122–136. [Waring's problem asks whether for every natural number k there exists an associated positive integer s such that k is the sum of at most s kth powers of natural numbers. For its history, see Vaughn and Wooley 2002; for a simplified account of Hilbert's solution, see Ellison 1971.]

[19] [Desargues's Theorem in projective geometry states that two triangles are in perspective axially if and only if they are in perspective centrally.]

[20] [For Klein's model, see Bonola 1955, 236–250.]

[21] Among later contributions to this question I mention W. Schwan, *Strecken-rechnung und Gruppentheorie*, *Math. Zeit.* vol. 3 (1919), pp. 11–28. A complete bibliography of geometric axiomatics since Hilbert would probably cover many pages. I refrain from citing a list of names.

[22] [For Helmholtz's arguments, see Pesic 2007, 47–52, and Helmholtz 1977, 39–71, discussed in modern notation in Adler, Bazin, and Schiffer 1975, 10–16.]

[23] A parallel development, with E. H. Moore as the chief prompter, must have taken place in this country. As I have to write from memory mainly, it is inevitable that my account should be colored by the local Göttingen tradition.

[24] *Math. Zeit.* vol. 10 (1921) [Weyl 1921b].

[25] [Weyl later uses this as an example of a false proposition; see below, 133.]

[26] [Zermelo's axiom of choice (1904) states, informally, that given any collection of sets, each set containing at least one member, it is possible to select exactly one member from each set. This axiom became controversial when it began to be applied to infinite collections of sets.]

[27] If we wish the rule that ρ_x binds x in all that comes after to be taken literally, we must write $\mathfrak{a} \to \mathfrak{b}$ in the form $\left\{ \begin{smallmatrix} \mathfrak{a} \\ \mathfrak{b} \end{smallmatrix} \right.$ The formulas will then look like genealogical trees.

[28] G. Gentzen, *Math. Ann.* vol. 112 (1936), pp. 493–565 [translated in Gentzen 1969].

[29] For later literature and systems of differential equations see Axel Schur, *Math. Ann.* vol. 82 (1921), pp. 213–239; G. A. Bliss, *Trans. Amer. Math. Soc.* vol. 28 (1926), pp. 561–584; W. T. Reid, ibid. vol. 44 (1938), pp. 508–521.

[30] H. Weyl and F. Peter, *Math. Ann.* vol. 97 (1927), pp. 737–755. A. Haar,

Ann. of Math. vol. 34 (1933), pp. 147–169. J. von Neumann, *Trans. Amer. Math. Soc.* vol. 36 (1934), pp. 445–492. Cf. also L. Pontrjagin, *Topological Groups*, Princeton, 1939. [Pontrjagin 1946]

[31] W. V. D. Hodge, *The Theory and Applications of Harmonic Integrals*, Cambridge, 1941. H. Weyl, *Ann. of Math.* vol. 44 (1943), pp. 1–6 [*WGA* 4:115–120].

[32] A book by Courant on the principle is in preparation [Courant 1950].

Mathematics and Logic
(1946)

1. Reduction of mathematics to set theory: the logical apparatus.

The reduction of mathematics to set-theory was the achievement of the epoch of Dedekind, Frege and Cantor, roughly between 1870 and 1895. As to the basic notion of set (to which that of function is essentially equivalent) there are two conflicting views: a set is considered either a collection of things (Cantor), or synonymous with a property (attribute, predicate) of things. In the latter case "x is a member of the set γ," in formula $x \in \gamma$, means nothing but that x has the property γ. The property of being red or being odd is certainly prior to the set of all red bodies or of all odd numbers. On the other hand, if with regard to a bag of potatoes or a curve drawn by pencil on paper, the property of a potato to be in the bag or of a point to lie on the curve is introduced, then the set (or a more concrete structure representing the set) is prior to the property. Whatever the epistemological significance of this distinction, it leaves the mathematician cool, since for any property γ we may speak of the set γ of all elements which have the property, and with respect to a given set γ we may speak of the property to be a member of γ. When he adopts the term *set* in preference to *property* the mathematician indicates his intention to consider co-extensive properties as identical, two properties α and β being co-extensive if every element that has the property α has also the property β and vice versa (set = "*Begriffsumfang* [extension of concept]," Frege).[1] Thus he will identify red and round in spite of their different "meanings" if every red body in the world happens to be round and vice versa.

The property π of "being prime" is represented by the propositional function $P(x)$, read "x is prime," with an argument x the range of significance of which is circumscribed by the concept "number." (The natural numbers $1, 2, 3, \cdots$ shall simply be called numbers; other numbers will be specified as rational, real, etc.) Indeed, understanding of the (false) proposition "6 is prime" requires that one understand what it means for any number x to be prime. Hence the proposition $P(6)$ arises from the propositional function $P(x)$ by the substitution $x = 6$. Besides properties we must consider binary, ternary, $\ldots$, relations, represented by propositional functions of 2, 3, ... arguments.

Although the mathematician need care little whether the language of properties or sets is used, he cannot afford to ignore another distinction sometimes confused with it: the distinction between what is considered as *given* and what he *constructs* from the given by the iterated combination of certain explicitly described constructive processes. For instance, in the axiomatic setup of elementary geometry one considers as given three categories of objects — points, lines (= straight lines) and planes, and a few relations between these objects (as "point *lies* on plane"). More complicated relations must be "defined," i.e., logically constructed from these primitive relations. In that intuitive theory of natural numbers (arithmetic) which is truly basic for all mathematics, even the objects are not given but constructed from the first number 1 by iterating one process, addition of 1; while all arithmetical relations are logically constructed from the one basic relation thus established: $y = x + 1$, "x is followed by y. " On the other hand, in a phenomenology of nature one would have to deal not only with categories of objects, as "bodies" or "events," but also with whole categories of properties which are prior to all construction, e.g. with the continuum of color qualities.

Logical construction of propositional functions from other propositional functions consists in the combined iterated application of a few elementary operations. Among them are the primitive logical operators $\sim$ (not), $\cap$ (and), $\cup$ (or), and the two quantifiers $(\exists x)$, "there exists an x such that," and $(\forall x)$, "for all x." For instance, from two propositional functions $S(x), T(x)$ we form

$$\sim S(x), \quad S(x) \cap T(x), \quad S(x) \cup T(x), \quad (\exists x)S(x), \quad (\forall x)S(x).$$

The quantifiers carry an argument x as index and "kill" that argument in the propositional function following the quantifier, just as the substitution of an individual number, say $x = 6$, does. The arithmetic operations $+$ and $\times$ are primarily applied to numbers and from there carry over to functions, while the process of integration with respect to a variable x by its very nature refers to a function $f(x)$. Just so, the logical operations $\sim, \cap, \cup$ deal primarily with propositions, while $(\exists x), (\forall x)$ refer to propositional functions involving a variable x. The operator $\cup$ is primitive in the sense that the truth value (true or false) of $a \cup b$ depends only on the truth values of a and b. The same holds for $\sim$ and $\cap$. It is convenient to add the primitive operator of implication for which I use Hilbert's symbol $\rightarrow$: $a \rightarrow b$ is false if a is true and b false, but true for the three other combinations: a true, b true; a false, b true; a false, b false. Propositions without argument result when all variables have been eliminated by substitution of explicitly given individuals or by quantifiers. The construction of arithmetic propositions and propositional functions constitutes their "meanings."

In introducing properties of numbers we presuppose that we know what we mean by "*any* number"; we shall say that the *category* of the elements to which the argument, in the propositional function under consideration refers must be given. We are assuming that this category is a closed realm of things existing in themselves, or, as we shall briefly say, is *existential*, when we ask with respect to a given property γ of its elements whether there *exists* an element of the property γ, with the expectation that, whatever the property γ, the question has a definite meaning and that there either exists such an element or every one of its elements has the opposite property $\sim \gamma$. In number theory or in elementary geometry we assume the numbers or the points, lines, planes, to constitute existential categories in this sense. However, we envisage only single individual properties and relations, never anything like the category of "all possible properties of numbers." This situation changes radically with the set-theoretic approach.

There we are forced to consider properties of numbers x as *objects* ξ of a new type to which the numbers stand in the relation $x \in \xi$. The proposition "6 is prime" is now looked upon as arising from the binary relation $x \in \xi$ by substituting 6 for x, and the property π of being prime for the argument ξ. The "copula" $\in$ corresponds to the word "is" in the spoken sentence "6 is prime." Any propositional function $P(x)$, like "x is prime," gives rise to a corresponding property $\pi = [x]P(x)$ (the property of being prime) such that $P(x)$ is equivalent to $x \in \pi$. The operator $[x]$ which effects transition from the propositional function to the corresponding property or set kills the argument x. For the sake of uniformity of notation we shall write henceforth $\in(x;\xi)$ instead of $x \in \xi$. In the same way a binary propositional function $P(x,y)$ defines a relation $\pi = [x,y]P(x,y)$ and $\in(xy;\pi)$ expresses the same as $P(x,y)$, namely that x and y are in this relation π.

2. Two examples.

Let our further reflections be guided by two typical examples taken from Dedekind's set-theoretic analysis of the two decisive steps in the building up of mathematics: his analysis of the sequence of numbers (in *Was sind und was sollen die Zahlen* 1887) and of the continuum of real numbers (in *Stetigkeit und Irrationalzahlen* 1872).[2] Our critical attention will be kept more alert by using Frege's terminology of properties rather than that of sets.

I. A property α of numbers is said to be hereditary if for any number x that has the property α, the follower $x + 1$ also has it. Dedekind defines: A number b is less than a if there *exists* a hereditary property that a has but b has not.

Here it is not only supposed that we know what we mean by *any* property, but we refer to the totality of all possible properties. In applying the quantifiers to properties of numbers as well as to numbers, it is absolutely imperative to look upon properties as secondary objects related to our primary objects, the numbers, by the copula relation $\in$. Heredity is even a property of properties. To be consistent we have to imagine objects of type 1 (the numbers), of type 2 (the properties of numbers), type 3, ..., and the fundamental relation $\in(x_n; x_{n+1})$ connects a variable x_n of type n with one x_{n+1} of type $n + 1$. Let $I(\xi)$ denote the proposition that ξ is hereditary. The definition of this propositional function whose Greek argument refers to the category "properties (or sets) of numbers" is as follows:

$$I(\xi) = (\forall x)\{\in(x; \xi) \to \in(x+1; \xi)\},$$

and Dedekind's definition of $x < y$:

$$(x < y) = (\exists \xi)\{I(\xi) \cap \in(y; \xi) \cap \sim\in(x; \xi)\}.$$

   II. Dedekind, as Eudoxus had done more than 2000 years before him, characterizes a non-negative real number α by the set of all positive rational numbers (fractions) $x > \alpha$. But for him any arbitrarily constructed (non-empty) set α of fractions (satisfying a certain condition, namely that along with any fraction b it contains every fraction $> b$) creates a corresponding real number α. The real number α is but a *façon de parler* for this set α. A set α consists of all fractions x satisfying a certain propositional function $A(x); \alpha = [x]A(x)$. Let $I(\xi)$ be a propositional function whose argument ξ refers to properties of fractions. The (greatest) lower bound $\gamma = [x]C(x)$ of a set of non-negative real numbers ξ can then be obtained as the *join* of all sets ξ for which $I(\xi)$ holds:

(1) $C(x) = (\exists \xi)\{I(\xi) \cap (x \in \xi)\}.$

In this way Dedekind *proves* that any set of non-negative real numbers has a lower bound. Again the quantifier $(\exists \xi)$ is applied to "all possible properties of fractions."

3. Levels or no levels? The constructive and the axiomatic standpoints.

But now let us pause to think what we have been doing. Properties of fractions are constructed by combined iterated application of a number of elementary logical operations $O_1, O_2, \cdots, O_h$. Let us call any property α thus obtained a *constructible* property, or a property of level 1. We can then interpret $(\exists \xi)$ in the definition (1) as "There exists a *constructible* property ξ," and this application of the quantifier is legitimate provided we admit its applicability to natural numbers. Indeed, the different manners in which one can form finite sequences with iteration out of h symbols $O_1, O_2, \cdots, O_h$ is not essentially more complicated than the possible finite sequences of one symbol 1. But the property $\gamma = [x]C(x)$ defined by (1) is certainly not identical in its meaning with any of the properties of level 1 because it is defined *in terms of the totality of all properties of level 1*. It is therefore a property of higher level 2. Nevertheless it may be *coextensive* with a property of level 1, and this is exactly what Russell's "axiom of reducibility" claims. But if the properties are constructed there is no room for an axiom here; it is a question which ought to be decided on the ground of the construction; and in our case that is a hopeless business. On the other hand, the edifice of our classical analysis collapses if we have to admit different levels of real numbers such that a real number is of level $l+1$ when it is defined in terms of the totality of real numbers of level l. If we wish to save Dedekind's proof we must abandon the constructive standpoint and assume that there is *given*, independently of all construction, an existential category "properties" or "second-type objects" (among which the constructible ones form but a small part) such that the following axiom holds, replacing the definition (1): For a given third-type object **i** there exists a second-type object γ such that

$$x \in \gamma \cdot \equiv \cdot (\exists \xi)\{(x \in \xi) \cap (\xi \in \mathbf{i})\}.$$

Here the arguments x and ξ refer to objects of first and second types, and $\equiv$ means "coextensive." That is a bold, an almost fantastic axiom; there is little justification for it in the real world in which we live, and none at all in the evidence on which our mind bases its constructions. With the assumption that properties constitute an existential category of given objects we return from Dedekind, who wanted to construct the real numbers out of the rational ones, to Eudoxus, for whom they were *given* by the points on a line; and instead of proving the existence of the lower bound on the ground of a definition of real numbers, we accept it as an axiom.

In reflecting on the source of the antinomies which had shown up at the fringe of Cantor's general set-theory, Russell realized the necessity of distinguishing between the several levels.[3] No doubt; by this fundamental insight, which he expressed somewhat loosely by his vicious circle principle: "No totality can contain members defined in terms of itself," he cured the disease but, as shown by the Dedekind example, also imperiled the very life of the patient. Classical analysis, the mathematics of real variables as we know it and as it is applied in geometry and physics, has simply no use for a continuum of real numbers of different levels. With his axiom of reducibility Russell therefore abandoned the road of logical analysis and turned from the constructive to the existential-axiomatic standpoint,—a complete volte-face.[4] After thus abolishing the several levels of properties, he still has the hierarchy of types: primary objects, their properties, properties of their properties, etc. And he finds that this alone will stop the known antinomies. But in the resulting system mathematics is no longer founded on logic, but on a sort of logician's paradise, a universe endowed with an "ultimate furniture" of rather complex structure and governed by quite a number of sweeping axioms of closure. The motives are clear, but belief in this transcendental world taxes the strength of our faith hardly less than the doctrines of the early Fathers of the Church or of the scholastic philosophers of the Middle Ages.

4. The Russell universe.

Let us describe this structure in a little more detail. A few primary categories of elements are given; they are the ranges of significance for the lowest types of arguments. In a relation, each of the n ($= 1$ or 2 or 3 or $\cdots$) arguments $x_1, \cdots, x_n$ refers to a certain "type" $k_1, \cdots, k_n$; the relation itself is of a type $k^* = \{k_1, \cdots, k_n\}$, determined by $k_1, \cdots, k_n$, that stands higher than any of the constituents $k_1, \cdots, k_n$. Draw a diagram representing k^* by a dot and $k_1, \cdots, k_n$ by a row of dots below k^* joined with it by straight lines, as you would depict a man k^* and his descendants in a geneological tree. Descending from k^* to its n constituents, and from them to their constituents, etc., one obtains a "topological tree" in which each *end point* is associated with one of the primary categories: this diagram describes the type k^*. The fundamental relation is $\in(x_1 \cdots x_n; x^*)$ where $x_1, \cdots, x_n$, refer to given types $k_1, \cdots, k_n$ and x^* to $k^* = \{k_1, \cdots, k_n\}$. Existential categories of elements are supposed to be given, one for each possible type (including the lowest, the primary types). We mentioned at the beginning with what data axiomatic elementary geometry operates, or a phenomenology of nature may have to operate; our present "Russell universe" U is seen to be incomparably richer.

We give a few of the more obvious axioms on which its theory is to be erected. The universal normal form for propositional functions involving n variables $x_1, \cdots, x_n$ of given types $k_1, \cdots, k_n$ is $\in(x_1 \cdots x_n; a^*)$. Indeed, such a relation is itself an element a^* of type $k^* = \{k_1, \cdots, k_n\}$. The relation of identity $(x = y)$ between elements of type k is itself an element of type $\{k, k\}$; call it $I = I_k$. Similarly, let $E = E_{k_1 \cdots k_n}$ be the $\in$ relation with its $n + 1$ arguments of types $k_1, \cdots, k_n, k^* = \{k_1, \cdots, k_n\}$. The existence of these special elements must be stipulated explicitly:

Axiom 1. There is an element $I = I_k$ of type $\{k, k\}$ such that $\in(xy; I)$ holds if and only if the elements x, y of type k are identical. There is an element $E = E_{k_1 \cdots k_n}$ of type $K = \{k_1, \cdots, k_n, k^*\}$ such that $\in(x_1 \cdots x_n x^*; E)$ is coextensive with $\in(x_1 \cdots x_n; x^*)$, the variables $x_1, \cdots, x_n$, and x^* ranging over their respective categories $k_1, \cdots, k_n$ and $k^* = \{k_1, \cdots, k_n\}$.

The composite property "red or round" is no longer constructed from the descriptive properties "red," "round," but belongs with them to the category of properties given prior to all construction. Its existence must be guaranteed by one of the simpler axioms of closure.

Axiom 2. Given an element a^* and an element b^* of type $k^* = \{k_1, \cdots, k_n\}$ there exists an element c^* of the same type such that $\in(x_1 \cdots x_n; c*)$ is coextensive with

$$\in(x_1 \cdots x_n; a^*) \cup \in(x_1 \cdots x_n; b^*)$$

(each x_i varying over its category k_i).

Substitution of a definite element b for a variable is taken care of by the

Axiom 3. Given an element a^* of type $k^* = \{k_1, \cdots, k_n\}$ and an element b_n of type k_n, there exists an element c of type $k = \{k_1, \cdots, k_{n-1}\}$ such that $\in(x_1 \cdots x_{n-1}; a)$ is coextensive with $\in(x_1 \cdots x_{n-1} b_n; a^*)$ (if $x_1, \cdots, x_{n-1}$ vary over the categories $k_1, \cdots, k_{n-1}$).

Elimination of a variable x_n, by the corresponding quantifier $(\exists x_n)$ changes a relation a^* of type k^* into a relation a of type k:

Axiom 4. Given an element a^* of type $k^* = \{k_1, \cdots, k_n\}$, there exists an element a of type $k = \{k_1, \cdots, k_{n-1}\}$ such that $\in (x_1 \cdots x_{n-1}; a)$ is coextensive with $(\exists x_n) \in (x_1 \cdots x_{n-1} x_n; a^*)$.

These are only a few typical axioms that indicate the direction. The reader must not look for them in the *Principia Mathematica*, which are conceived in a different style. But our system U embodies the same ideas in a form that seems to me both natural in itself and advantageous for a comparison with other systems W, Z presently to be discussed.

Our axioms serve as a basis for deduction in the same way as, for instance, the axioms of geometry; deduction takes place by that sort of logic on which one is used to rely in geometry or analysis, including the free use of "there exists" and "all" with reference to any fixed type in the hierarchy of types an dthe elements of the corresponding categories. While these categories, as well as the basic relation $\in$, are considered as undefined, the logical terms like "not" $\sim$, "if then" $\to$, "there is" $(\exists x)$, etc., have to be understood in their meaning and do not form part of the axiomatic system: the formalism of symbolic logic is merely used for the sake of conciseness.

If the *Principia Mathematica* set out to base mathematics on pure logic the result, as we now see, is quite different: an axiomatic world system has taken the place of logic. Its very structure, the hierarchy of types, cannot be described without resort to the intuitive concept of iteration. To develop, in Dedekind-Frege fashion, a theory of natural numbers from this system is therefore an enterprise of doubtful value.

5. A constructive compromise.

Realizing the highly transcendental character of the axiomatic universe from which this system deduces mathematics, one wonders whether it is not possible to stick, in spite of everything, to the constructive standpoint, which seems so much more natural to the mathematician. We accept the hierarchy of types; but we assume only one category of primary objects, the numbers; and one basic binary relation between numbers, namely "x is followed by y." All other relations of the various types are explicitly constructed, the quantifiers $(\exists x)$ and $(\forall x)$ being applied only to numbers and not to arguments of higher type. No axioms are postulated. What we can get in this way constitutes the ground level, or level 1. One could build over it a second level containing relations which are constructed by applying the quantifiers to the totality of relations of this or that type constructible on the first level, and proceed in the same manner from the second to a third level, etc. One would obtain a "ramified hierarchy" of types and levels. But in this way, as we have said before, nothing resembling our classical Calculus will result. The temptation to pass beyond the first level of construction must be resisted; instead, one should try to make the range of constructible relations as wide as possible *by enlarging the stock of basic operations.* It is a priori clear that *iteration* in some form must find a place among these irreducible principles of construction—contrary to the Dedekind-Frege program.

We begin again at the beginning. Let $R(x, \xi)$ be a binary propositional function the two arguments x, ξ of which are of types k, κ respectively; for instance the relation "x is less than ξ" between numbers. We can then speak of the property of a number "to be less than ξ" (or of the set of all numbers less than ξ). This is obviously a property depending on ξ. In general terms we can form $[x]R(x, \xi) = r^*(\xi)$, which is an element of $\{k\}$ depending on ξ such that $R(x, \xi)$ is coextensive with $\in(x; r^*(\xi))$. If in addition to $R(x, \xi)$ we have a propositional function $S(x^*)$ whose argument is of type $k^* = \{k\}$, we can form $T(\xi) = S(r^*(\xi))$, or more explicitly,

$$T(\xi) = S([x]R(x, \xi)).$$

This is the *process of substitution* generating $T(\xi)$ from $R(x, \xi)$ and $S(x^*)$.

Take now the particular situation that $\kappa = \{k\}$. Then argument and value of the function $r*(\xi) = [x]R(x, \xi)$ are of the same type κ, and whenever that happens *iteration* becomes possible. Thus we define by complete induction a relation $T(n, \xi)$ in which the argument n refers to the primary category of numbers, as follows:

$$T(1, \xi) = S(\xi);$$

$$T(n + 1, \xi) = T(n, r^*(\xi)) \qquad (n = 1, 2, \cdots).$$

Adding the operation of substitution and iteration, as illustrated by these examples, to the other elementary logic operations, but without applying the quantifiers to anything else than numbers, the writer was able (in *Das Kontinuum*, 1918) to build up in a purely constructive way, and without axioms, a fair part of classical analysis, including for instance Cauchy's criterion of convergence for infinite sequences of real numbers.[5] In this system iteration plays the role which in set-theory was played by the uninhibited application of quantifiers. Our construction honestly draws the consequences of Russell's logical insight into the tower of levels, which Dedekind had ignored inadvertently and Russell himself, afraid of its consequences, razed to the ground by his axiom of reducibility. Considering their common origin the axiomatic system U as outlined in section 4 and this constructive approach are remarkably different. But even here we have adhered to the belief that "there is" and "all" make sense when applied to natural numbers: in addition to logic we rely on this existential creed and the idea of iteration.

6. Brouwer's intuitionistic mathematics.

Essentially more radical and a further step toward pure constructivism is Brouwer's intuitionistic mathematics.[6] Brouwer made it clear, as I think beyond any doubt, that there is no evidence supporting the belief in the existential character of the totality of all natural numbers, and hence the principle of excluded middle in the form "Either there is a number of the given property γ, or all numbers have the property $\sim \gamma$" is without foundation. The first part of the sentence is an *abstract* from some statement of fact in the form: The number thus and thus constructed has the property γ. The second part is one of *hypothetic* generality, asserting something only if $\cdots$; viz., if you are actually given a number, you may be sure that it has the property $\sim \gamma$. The sequence of numbers which grows beyond any stage already reached by passing to the next number, is a manifold of possibilities open towards infinity; it remains forever in the status of creation, but is not a closed realm of things existing in themselves. That we blindly converted one into the other is the true source of our difficulties, including the antinomies—a source of more fundamental nature than Russell's vicious circle principle indicated. Brouwer opened our eyes and made us see how far classical mathematics, nourished by a belief in the "absolute" that transcends all human possibilities of realization, goes beyond such statements as can claim real meaning and truth founded on evidence. According to his view and reading of history, classical logic was abstracted from the mathematics of finite sets and their subsets. (The word finite is here to be taken in the precise sense that the members of such a set are explicitly exhibited one by one.) Forgetful of this limited origin, one afterwards mistook that logic for something above and prior to all mathematics, and finally applied it, without justification, to the mathematics of infinite sets. This is the Fall and original sin of set-theory, for which it is justly punished by the antinomies. Not that such contradictions showed up is surprising, but that they showed up at such a late stage of the game!

Thanks to the notion of "*Wahlfolge* [choice sequence]," that is a sequence *in statu nascendi* in which one number after the other is *freely chosen* rather than determined by law, Brouwer's treatment of real variables is in the closest harmony with the intuitive nature of the continuum; this is one of the most attractive features of his theory.[7] But on the whole, Brouwer's mathematics is less simple and much more limited in power than our familiar "existential" mathematics. It is for this reason that the vast majority of mathematicians hesitate to go along with his radical reform.

7. The Zermelo brand of axiomatics; sets and classes.

From this excursion to the left wing of the "constructionists" we return to the universe U with its hierarchy of types. Once one has committed oneself to the existential or axiomatic viewpoint, can one not go forth in the same direction and even erase all differences of types, taking only such precautions as are absolutely necessary to avoid the known contradictions? This is what Zermelo did in his *Untersuchungen über die Grundlagen der Mengenlehre* [*Investigations on the Foundations of the Theory of Sets*], 1908.[8] His axioms deal with only one (existential) category of objects called elements or sets, and one basic relation $x \in y$, "x is member of y." But he is forced to give up the principle that any well-defined property γ determines an element c such that $x \in c$ whenever the element x has the property γ and vice versa. Properties are used by him merely to cut out subsets *from a given set.* Hence his axiom of selection: "Given a well-defined property γ and an element a, there is an element a' such that $x \in a'$ if and only if x is member of a and has at the same time the property γ." The notion of a well-defined property which enters into it is somewhat vague. But we know that we can make it precise by constructing properties by iterated combined application of some elementary constructive processes. Instead of saying that x has the property γ, let us say that x is a member of the class $\gamma, x \in \gamma$. We thus distinguish between elements or sets on the one hand, classes on the other, and formulate the axioms in terms of two undefined categories of objects, elements and classes. Since we postulate that two elements a, b are identical in case $x \in a$ and $x \in b$ are coextensive, and since each element a is associated with the class α of all elements x satisfying the condition $x \in a$, we are justified in identifying a with that class α. Then *every element is a class* and the axioms deal with one undefined fundamental relation $x \in \xi$, "the element x is member of the class ξ," which has absorbed Zermelo's relation $x \in y$ between elements. The principles for the construction of properties are replaced by corresponding axioms for classes; e.g., given two classes α and β, there exists a class γ such that the statement $(x \in \alpha) \cup (x \in \beta)$ about an arbitrary element x is coextensive with $x \in y$.

Since the axiom of selection can only generate smaller sets out of a given set, we need some vehicle that carries us in the opposite direction. Therefore two axioms are added guaranteeing the existence of the set of all subsets of a given set and the join of a given set of sets. It is essential that they be limited to sets = elements and do not apply to classes.

With the introduction of classes, which is due to Fraenkel, von Neumann, Bernays, and others, the axioms assume the same self-sustaining character as, for instance, the axioms of geometry; no longer do such general notions as "any well-defined property" penetrate into the axiomatic system from the outside. A complete table of axioms for this system, Z as we shall call it, is to be found on the first pages of Gödel's monograph, *Consistency of the Continuum Hypothesis.*[9] Even before the turn of the century Cantor himself had moved in the same direction by distinguishing "consistent classes" = sets and inconsistent classes.[10] Not the hierarchy of types, but the non-admission among the decent "sets" of such classes as are too "big," averts here the disaster of the familiar antinomies.

One might object to a system like Z on the ground that it does not rest on a real insight into the causes of the antinomies but patches up Cantor's original conception by a minimum of concessions necessary to avoid the known contradictions. Indeed, we have no assurance of the consistency of Z, except from the empirical fact that so far no contradictions have resulted from it. But we are in no better position toward Russell's universe U. And Z has its great advantages over U: it is of an essentially simpler structure and seems to be the most adequate basis for what is actually done in present-day mathematics. In particular, the "existential" Dedekind-Frege theory of numbers can be derived from it (Zermelo), and Gödel was able to show that Zermelo's far-reaching axiom of choice in a very sharp form is consistent with the other axioms of Z.[11]

8. Complete formalization and the question of consistency. Pessimistic conclusions.

A new turn in the axiomatization of mathematics of paramount importance was inaugurated by Hilbert's "*Beweistheorie*" [theory of proof] (since 1922).[12] Hilbert sets out *to prove* (not the truth, but) *the consistency of mathematics*. He realizes that to that end mathematics and logic must first be completely formalized: all statements are to be replaced by formulas in which now also the logical operators $\sim, \cap, (\exists x)$ etc., appear as undefined symbols. Thus formalized logic is absorbed into formalized mathematics.[13] The formulas have no meaning. A mathematical demonstration is a concrete sequence of formulas in which a formula is derived from the preceding ones according to certain rules comprehensible without recourse to any meaning of the formulas — just as in a game of chess each position is derived from the preceding one by a move obeying certain rules. *Consistency*, the fact that no such game of deduction may end with the formula $\sim(1 = 1)$, must be proved by intuitive reasoning about the formulas which rests on evidence rather than on axioms, and respects throughout the limits of evidence as disclosed by Brouwer. But in this thinking about demonstrations, in following a hypothetic sequence of formulas leading up to the end formula $\sim(1 = 1)$ our mind cannot help using that type of evidence in which the possibility of iteration is founded. In the axiomatization of mathematics Hilbert is forced to proceed with more restraint than Zermelo: if he is too liberal with his axioms he will lose all chance of ever proving their consistency; he is guided by an at least vaguely preconceived plan for such a proof. It is for this reason that he finds it advisable, for instance, to distinguish various levels of variables.

Hilbert's formulas are concrete structures consisting of concrete symbols; the order in which the symbols follow one another in a formula, and also their identity in the same or different formulas, must be recognizable irrespective of little variations in their execution. Handling these symbols, we move on the same level of understanding as guides our daily life in our relationship to such tools as hammer or table or chair. Hilbert sees in it the most important prelogical foundation of mathematics, in fact of all science. But in addition his axioms of mathematics and the intuition of iteration of which the metamathematical nonaxiomatic reasoning about mathematics makes use are other extralogical ingredients of his system.

Our brief survey may be summarized in a little diagram in which the con-

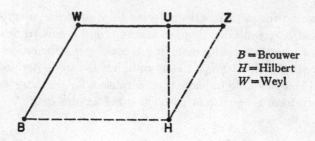

structive tendency increases toward the left, the axiomatic toward the right, and also the relative "depth" of the foundations is indicated. Frege and after him Russell had hoped (1) to develop the theory of natural numbers on a sure basis without recourse to the intuitive idea of indefinite iteration, and (2) to make mathematics a part of logic. We have now seen that none of the systems discussed gives any hope of accomplishing (2), the subjugation of mathematics by logic. In U and Z elaborate systems of axioms form the basis, in W, B, and to a lesser degree also in U, the intuitive idea of iteration is indispensable. The extra-logical foundations of Hilbert's theory have just been described. The only system which in a sense can claim to reach the goal (1) is Z. But even there the theory of numbers does not rest on logic alone, but on a highly transcendental system of axioms (the belief in whose consistency is supported by empirical facts, but not by reasons). Poincaré has proved right in his defense of mathematical induction as an indispensable and irreducible tool of mathematical reasoning.

It is likely that all mathematicians ultimately would have accepted Hilbert's approach had he been able to carry it out successfully. The first steps were inspiring and promising. But then Gödel dealt it a terrific blow (1931), from which it has not yet recovered. Gödel enumerated the symbols, formulas, and sequences of formulas in Hilbert's formalism in a certain way, and thus transformed the assertion of consistency into an arithmetic proposition. He could show that this proposition can neither be proved nor disproved within the formalism.[14] This can mean only two things: either the reasoning by which a proof of consistency is given must contain some argument that has no formal counterpart within the system, i.e., we have not succeeded in completely formalizing the procedure of mathematical induction; or hope for a strictly "finitistic" proof of consistency must be given up altogether. When G. Gentzen finally succeeded in proving the consistency of arithmetic he trespassed those limits indeed by claiming as evident a type of reasoning that penetrates into Cantor's "second class of ordinal numbers."[15]

From this history one thing should be clear: we are less certain than ever about the ultimate foundations of (logic and) mathematics. Like everybody and everything in the world today, we have our "crisis." We have had it for nearly fifty years. Outwardly it does not seem to hamper our daily work, and yet I for one confess that it has had a considerable practical influence on my mathematical life: it directed my interests to fields I considered relatively "safe," and has been a constant drain on the enthusiasm and determination with which I pursued my research work. This experience is probably shared by other mathematicians who are not indifferent to what their scientific endeavors mean in the context of man's whole caring and knowing, suffering and creative existence in the world.

Notes

[Published with the subtitle "a brief survey serving as a preface to a review of *The Philosophy of Bertrand Russell*" in Weyl 1946a; *WGA* 4:268–279. Weyl 1946b gives the text of his review of Schilpp 1951. In its original printed version, Weyl used the printed Greek symbol ϵ rather than its common present orthography as $\in$, which we have used throughout the text here. Also, Weyl's original references have been included as notes below.]

[1] [See Burge 1984 for a discussion of this term in context.]

[2] [Dedekind's two fundamental writings, translated as "The Nature and Meaning of Numbers" and "Continuity and Irrational Numbers" are available in Dedekind 1963 and in Ewald 1996, 2:787–832, 765–778.]

[3] "Mathematical Logic as Based on the Theory of Type," *Am. Jour. Math.*, v. 30, 1908, pp. 222–262. B. Russell and A. N. Whitehead, *Principia Mathematica*, 3 vols., Cambridge, 1910–13; 2nd ed. of vol. I, 1935.

[4] I know very well that this is at odds with Russell's own interpretation; he in the course of time became more and more inclined to visualize sets as "logical fictions." "Though," as Gödel adds, "perhaps the word fiction need not necessarily mean that these things do not exist, but only that we have no direct perceptions of them." In the second edition of Volume I of the *Principia Mathematica*, an attempt is made to prove independently of the axiom of reducibility that at least all levels of *natural numbers* can be reduced to the five lowest. But as Gödel observes, the proof is far from being conclusive. [See] Gödel, "Russell's Mathematical Logic," in *The Philosophy of Bertrand Russell*, pp. 127, 145, 146 [Schilpp 1951].

[5] But of course the theorem of the lower bound of an arbitrary set of non-negative real numbers could not be upheld. [For a translation of *Das Kontinuum*, see Weyl 1994.]

[6] Brouwer's thesis "Over de grondslagen der wiskunde" appeared in 1907 [translated in Brouwer 1975]. For a list of his papers on the subject see A. Church's "Bibliography of Symbolic Logic," *Jour. Symb. Logic*, v. 1, 1936, pp. 121–218. [For helpful selections of his texts, see Ewald 1996, 2:1166–1207 and Mancosu 1998, 1–64, 286–295.]

[7] [See Van Atten 2007.]

[8] *Annals of Mathematics Studies* No. 3, Princeton, 1940; *Math. Ann.*, v. 65, 1908, pp. 261–281. [See also Ewald 1996, 2:1208–1233.]

[9] See also P. Bernays, *Jour. Symb. Logic*, v. 2, 1937, pp. 65–77 and the references given there. [Gödel 1986–2003, 2:33–101, reprints his monograph on the consistency of the continuum hypothesis.]

[10] Cf. G. Cantor, *Gesammelte Abhandlungen*, ed. E. Zermelo, 1932, pp. 443–451 (correspondence Cantor-Dedekind) [Cantor 1980].

[11] [See Gödel 1986–2003, 2:33–101.]

[12] Vol. 3 of Hilbert's *Collected Papers*, 1937, and Hilbert-Bernays, *Grundlagen der Mathematik*, 2 vols. 1934 and 1939 [Hilbert and Bernays 1968–70].

[13] In this regard the ground was well prepared for Hilbert by the *Principia Mathematica*. For the sake of comparison I mention one other completely formalized system, that of Quine. [See] *Am. Math. Monthly*, v. 44, 1937, pp. 70-80.

[14] *Monatsh. Math. Phys.*, v. 38, 1931, pp. 173–198 [Gödel 1986–2003, 1:144–195].

[15] *Math. Ann.*, v. 112, 1936, pp. 493–565 [in Gentzen 1969].

Relativity Theory as a Stimulus in Mathematical Research

(1949)

About the influence which the physical theory of relativity had upon purely mathematical research different mathematicians will be of different opinions. But it is unlikely that anybody today would agree with E. Study, Felix Klein's contemporary and life-long enemy, a man of considerable merit in his field and of violent temper, who, in a book published 1923, accused the writers on relativity theory and tensor calculus of having laid waste a rich cultural domain (*ein reiches Kulturgebiet der Verwahrlosung anheimgegeben zu haben*), that rich cultural field being the algebraic theory of invariants. What got Study's goat was the fact that the symbolic method and the classical notations of that theory had been more or less ignored by the relativists. I shall come back to his problems later on. Anyhow I would not stand here and try to say some words about the topic which the program announces, were I not convinced that relativity theory during the last decades has been an invigorating rather than a devastating influence in the development of several branches of mathematics, including the theory of invariants. Nor is it difficult to prove my point; the facts speak a too unmistakable language.

The relativity problem is one of central significance throughout geometry and algebra and has been recognized as such by the mathematicians at an early time. It played a great role in Leibniz's philosophical-mathematical ideas. In the nineteenth century the concept of a group of transformations was devised and developed as the adequate tool for dealing with it. Suppose a realm of objects, which may be called points, is given to you. Those transformations, those one-to-one mappings of the point field into itself which leave all relations of objective significance between points undisturbed, form a group, the group of automorphisms. If in some way coordinates, self-created reproducible symbols like numbers, are assigned to the points then any automorphism carries this assignment over into a new one from which it is objectively indiscernible (to use Leibniz's word). Hence any coordinate assignment requires an act of choice by which one picks out one from a class of equally admissible coordinate systems. The class is objectively characterized, but not the individual coordinate assignment. In Galois' theory the "points" are the n roots $\alpha_1, \cdots, \alpha_n$ of an algebraic equation of degree n with rational coefficients. The objective relations are those expressible in terms of the fundamental operations of addition, multiplication, subtraction, and division, i.e., all relations of the form $F(\alpha_1, \cdots, \alpha_n) = 0$ where $F(x_1, \cdots, x_n)$ is a polynomial of the n variables $x_1, \cdots, x_n$, with rational coefficients. Labeling the roots by $1, 2, \cdots, n$ amounts to a coordinate assignment. A transformation is a permutation of the n roots, an automorphism a permutation which leaves all algebraic relations $F(\alpha_1, \cdots, \alpha_n) = 0$ with rational coefficients undisturbed; the automorphisms form the Galois group. Examples from geometry are probably more familiar to this audience. In the three-dimensional Euclidean vector space a frame of reference consists of 3 mutually perpendicular vectors of length 1. Relative to such a frame any vector can be described by the triple of its coordinates $(x_1 x_2 x_3)$. Transition from one such frame to an arbitrary other is effected by an orthogonal transformation of the coordinates x_1, x_2, x_3, and these transformations constitute the group of automorphisms. In affine vector geometry this group is replaced by the group of all homogeneous linear transformations with non-vanishing determinant (or, if one follows Euler, with determinant 1). In the first half of the nineteenth century projective geometry had arisen, whose group of automorphisms consists of all collineations, i.e., of all point transformations that carry straight lines into straight lines. Möbius had added his spherical geometry, the automorphisms of which carry spheres into spheres. In a space of three or more dimensions the group of Möbius transformations coincides with that of all conformal transformations. Ideas which seem to have guided Möbius implicitly in his investigations, but could not be formulated without the group concept, were made explicit by Felix Klein in his famous Erlanger program, 1872.[1] Transitions between equally admissible coordinate assignments or frames of reference in a Klein space find their expression in a group Γ of coordinate transformations. Klein defines the geometry by this group, which the mathematician feels free to choose as he likes: point relations are then said to be of objective significance if they are invariant with respect to the group Γ, two configurations of points are considered objectively alike if one is carried into the other by an operation of the group. For instance, if the group is transitive, as we shall assume in the future, all points are alike, the space is homogeneous.

According to Einstein's special relativity theory the four-dimensional world of the space-time points is a Klein space characterized by a definite group Γ; and that group is the one most familiar to the geometers, namely the group of Euclidean similarities—with one very important difference however. The orthogonal transformations, i.e., the homogeneous linear transformations which leave

$$(+) \qquad\qquad x_1^2 + x_2^2 + x_3^2 + x_4^2$$

unchanged have to be replaced by the Lorentz transformations leaving

$$(-) \qquad\qquad x_1^2 + x_2^2 + x_3^2 - x_4^2$$

invariant. This was certainly a surprise to the mathematicians. But it did not disturb them very greatly; Minkowski made the necessary adjustments at once. Indeed in algebraic geometry they had become used to considering their variables as capable of arbitrary complex values; for that made their theories so much simpler, owing to the fact that the field of all complex numbers is algebraically closed, i.e., that an arbitrary algebraic equation of degree n with coefficients in the field always has n roots in the field. Now in the domain of complex numbers there is no difference between $(+)$ and $(-)$; indeed $(+)$ goes over into $(-)$ by substituting ix_4 for x_4. But in the last forty-odd years algebra has reversed its position: not only has it recognized the right of the field of real numbers besides that of the complex numbers, but it carries on its investigations, whenever possible, in an arbitrarily given field of numbers, no longer assuming that this underlying field, though closed with respect to the operations of addition, subtraction, multiplication, and division, be also algebraically closed. Two non-degenerate quadratic forms with coefficients in a given field $\mathfrak{F}$ belong to the same genus if one may be carried into the other by a linear transformation with coefficients in $\mathfrak{F}$. Special relativity could have taught the algebraists this lesson: do not ignore other genera of quadratic forms besides the principal genus represented by the unit form $x_1^2 + x_2^2 + \cdots$. As a matter of fact, in their arithmetical theory of quadratic forms, a classic subject from Gauss to Minkowski, they had never ignored them. I am afraid, the geometers had. Yet one can hardly say that here the mathematicians received a new stimulus from relativity theory; rather Klein's Erlanger program and the distinction of genera of quadratic forms was in happy concordance with relativity theory, and Einstein's discovery gave support to these geometric and algebraic conceptions by exhibiting one very important and quite unexpected application in physics.

We are used today to look at mathematical questions from the standpoints of abstract algebra and of topology. Between the orthogonal and the Lorentz group there is a topological difference much more incisive than the algebraic difference of genus: the one is a compact manifold, the other is not. The most systematic part of group theory deals with the representations of groups by linear transformations. Representations in a Hilbert space of finite or infinitely many dimensions are of supreme interest to quantum mechanics. If the group is finite every such representation breaks up into irreducible parts of finite dimensionality; the entire theory, one of the proudest buildings of mathematics, is dominated by the orthogonality and completeness relations. They carry over from finite to compact groups. The theory of Fourier series is nothing but the representation theory of the group of rotations of a circle. Pleased by the beauty and harmony of this theory of representations of compact groups, the mathematicians shyed away for some time from the more complicated and less harmonious situation that had to be expected for non-compact groups. But the Lorentz group and the interest which quantum mechanics has in its representations in Hilbert space finally forced the issue: V. Bargmann in this country and Gelfand and Neumark in Russia mustered enough courage to tackle the representations of the Lorentz group in Hilbert space, and the Russians went on to develop the theory for any groups that are locally but not globally compact.

There is hardly any doubt that for physics special relativity theory is of much greater consequence than the general theory.[2] The reverse situation prevails with respect to mathematics: there special relativity theory had comparatively little, general relativity theory very considerable, influence, above all upon the development of a general scheme for differential geometry. The kind of world geometry Einstein needed to put into mathematical form his central idea of an inertial field that not only acts upon matter but is also acted upon by matter, he found ready-made in the mathematical literature: from Gauss's theory of surfaces in Euclidean space Riemann had abstracted his conception of an n-dimensional Riemannian space. Here the coordinate assignment remains quite arbitrary, subject to arbitrary (differentiable) transformations. A coordinate system usually covers only a part of the manifold; a finite or even an infinite number of partially overlapping patches are needed to cover it completely. But this is of little concern to us, as long as we are still far from overlooking the four-dimensional world in its entire extension. The fact that in an infinitesimal neighborhood of a point Pythagoras' theorem is supposed to prevail finds its expression in Riemann's formula

$$(*) \qquad\qquad ds^2 = \sum_{i,j} g_{ij} dx_i dx_j \qquad (g_{ji} = g_{ij})$$

for the square of the length ds of a line element that leads from the point $P = (x_1, \cdots, x_n)$ to an arbitrary infinitely near point $P' = (x_1 + dx_1, \cdots, x_n + dx_n)$. The coefficients g_{ij} do not depend on the line element but will in general vary from point to point. Riemann had gone some distance in developing this kind of geometry, which clearly follows a trend entirely different from Klein's Erlanger program; he had, in particular, derived what is now called the Riemann curvature tensor. An elaborate mathematical machinery for Riemannian geometry had been set up under the name of Absolute Differential Calculus by G. Ricci and T. Levi-Civita. These things, which Einstein learnt from his friend, the mathematician Marcel Grossman in Zurich, enabled him to write down the equation of motion for a planet and the differential equations for the gravitational field, without an appeal to experience, in a purely speculative and yet astonishingly compelling manner. And Nature graciously confirmed his laws with as clear an O.K. as one can ever get from her.

The great importance which Riemannian geometry acquired for Einstein's theory of gravitation gave the impetus to develop this geometry further, to study more carefully its foundations and, as a consequence of such analysis, to generalize it in various directions. The first and decisive step was Levi-Civita's discovery of the notion of infinitesimal parallel vector displacement. Let us begin with Gauss' representation of a surface in Euclidean space. The points P of the surface are referred to arbitrary coordinates x_1, x_2, the location of the point P in the embedding space is given by $\mathfrak{r} = \mathfrak{r}(x_1 x_2)$ where $\mathfrak{r} = \overrightarrow{OP}$ is the vector leading in that space from the origin O to the space point P. The increment

$$d\mathfrak{r} = \frac{\partial \mathfrak{r}}{\partial x_1} dx_1 + \frac{\partial \mathfrak{r}}{\partial x_2} dx_2$$

is the line element joining $P = (x_1, x_2)$ with the nearby surface point $P' = (x_1 + dx_1, x_2 + dx_2)$. In this way it comes about that the two-dimensional linear manifold of the tangent vectors $\mathfrak{r}$ in P is referred to the affine vector basis consisting of the two vectors

$$\mathfrak{e}_1 = \partial \mathfrak{r}/\partial x_1, \mathfrak{e}_2 = \partial \mathfrak{r}/\partial x_2 : \ \mathfrak{r} = \xi_1 \mathfrak{e}_1 + \xi_2 \mathfrak{e}_2.$$

The metric structure of this vector compass is taken into account, only afterwards, as it were, by expressing the square of the length of $\mathfrak{r}$ as a quadratic form $\sum g_{ij} \xi_i \xi_j$ of its components with certain coefficients g_{ij}. Thus the two-dimensional Euclidean vector space, the compass at P, is here treated as an affine space to which a quadratic form is joined as the "absolute." In a similar fashion affine space is sometimes treated as projective space in which one plane has been absolutely distinguished as the "plane at infinity." There is an obvious artificiality in this procedure as it falsifies the group, so to speak; but it is also obvious how Gauss' approach led to this treatment. But the natural frame for the compass would be a Cartesian one consisting of two perpendicular vectors $\vec{e}_1, \vec{e}_2$ of length 1 at the point P; choose such a frame without tying it to the coordinates x_1, x_2 to which the neighborhood P on the surface is referred. The line element $\overrightarrow{PP'}$ will then be given by an expression $\overrightarrow{PP'} = \sum \omega_i \vec{e}_i$ where the ω_i are linear differential forms $o_{i1} dx_1 + o_{i2} dx_2$. Invariance must prevail (1) with respect to arbitrary transformations of the coordinates x_i, (2) with respect to an arbitrary rotation of the frame $\vec{e}_i$ of the vector compass at P, arbitrary in the sense that it may depend in an arbitrary fashion on P. This scheme which E. Cartan always used is better suited to the nature of the Pythagorean metric. When one tries to fit Dirac's theory of the electron into general relativity, it becomes imperative to adopt the Cartan method. For Dirac's four ψ-components are relative to a Cartesian (or rather a Lorentz) frame.[3] One knows how they transform under transition from one Lorentz frame to another (spin representation of the Lorentz group); but this law of transformation is of such a nature that it cannot be extended to arbitrary linear transformations mediating between affine frames.

Let us return to the surface in Euclidean space. A tangent vector $\mathfrak{r}$ at P can be transferred to the nearby surface point P' by parallel displacement in the embedding Euclidean space. The $\mathfrak{r}$ thus obtained is no longer tangential at P'; we therefore split it into its tangential and its (infinitesimal) normal component at $P', \mathfrak{r} = \mathfrak{r}' + \mathfrak{r}_n$, and throw away the latter: $\mathfrak{r} \rightarrow \mathfrak{r}'$ is Levi-Civita's process of infinitesimal parallel displacement on the surface. It first looks as if it depended on the embedment of the surface into the surrounding Euclidean space; but when one carries out the computation, it turns out to be completely determined by the intrinsic metric of the surface. Hence it must be possible to give an intrinsic definition of the process. The first obvious property of Levi-Civita's displacement is that it leaves the length of the vector $\mathfrak{r}$ unchanged. In combination with another intrinsic condition, which I shall not formulate but only allude to by the word "condition of closure," this leads indeed to a unique determination of the "affine connection" by virtue of which each vector at P goes over into a definite vector at the infinitely near point P'. After the notion is thus made independent of an embedding Euclidean space, it becomes applicable to any two-dimensional, nay to any n-dimensional Riemannian manifold.

It is natural to introduce the concept of an *affinely connected manifold* as one in which the process of infinitesimal parallel vector displacement satisfying the condition of closure is defined.[4] It is a fact that the whole tensor analysis with its "covariant derivatives" makes use of the affine connection only, not of the metric. Also Riemann's curvature finds its place here. Indeed if one carries the compass at P by successive steps of infinitesimal parallel displacement around a circuit returning to the start P, then the compass will in general not return to its initial position, but in one that arises from the initial position by a certain rotation around P. This rotation is essentially what Riemann called curvature and what should perhaps more appropriately be called vector vortex. The affine notions of covariant differentiation and of curvature become applicable to Riemannian geometry owing to the fact that the Riemann metric uniquely determines the affine connection.

Thus an affine infinitesimal geometry has sprung up beside Riemann's metric one. One may say that the causal structure of the universe in the immediate neighborhood of the world point P is described by the equation $\sum g_{ij} dx_i dx_j = 0$. Hence a Riemannian metric g_{ij} and a metric g_{ij}^* of the same space lead to the same causal or conformal structure if $g_{ij}^* = \lambda g_{ij}$ with a positive factor λ depending on P in an arbitrary fashion.[5] Such features of a Riemannian space are conformal as are not affected by the change $g_{ij} \to \lambda g_{ij}$. One can also easily describe under what conditions two affine connections are equivalent in the sense that they lead to the same geodesics and thus to the same inertial or projective structure. What happens, one may further ask, when one replaces the geodesics by any families of curves of such nature that through every point in every direction there goes one of these lines? (General geometry of paths.)

Now here is clearly rich food for mathematical research and ample opportunity for generalizations. Thus schools of differential geometers sprang up in the wake of general relativity. Here in Princeton Eisenhart and Veblen took the lead, Schouten in Holland. In France, E. Cartan's fertile geometric imagination disclosed many new aspects of the subject.[6] Some of their outstanding pupils are Tracy Thomas and J. M. Thomas in Princeton, van Dantzig in Holland and Shiing Shen Chern of the Paris school. A lone wolf in Zurich, Hermann Weyl, also busied himself in this field; unforunately he was all too prone to mix up his mathematics with physical and philosophical speculations. In several ways these authors soon arrived at the conclusion that it is better to establish projective differential geometry not by abstraction from the affine brand, as described above, but independently, namely by associating with each point P of the manifold a projective space Σ_P in the sense of Poncelet and Plücker, this homogeneous space taking the place of the affine vector compass in the affinely connected manifold. In the same manner a general conformal geometry may be developed by associating with each point P a Möbius space Σ_P. The generalization is evident. Let a manifold M and a homogeneous Klein space Σ, defined by a transitive group Γ of transformations, be given. Assume that with each point P there is associated a copy Σ_P of the Klein space, and that Σ_P is carried over to the space $\Sigma_{P'}$ associated with an infinitely near point P' of M by an infinitesimal operation of the group Γ that depends linearly on the relative coordinates dx_i of P' with respect to P. The manifold M, or at least part of it, is referred to coordinates x_i. In each Σ_P we must choose an admissible frame of reference, with respect to which the points of Σ_P are represented by coordinates ξ. Since the Klein space is supposed to generalize the affine tangent vector space, it is natural to assume that a definite center O in Σ_P is marked which "covers" the point P on M. The frame for Σ_P may be so chosen that O becomes the origin $\xi_1 = \xi_2 = \cdots = 0$. It is further natural to assume that the infinitesimal vectors issuing from O in Σ_P on the one hand, and those issuing from P in M on the other hand, are "in coverage" by dint of a one-to-one linear mapping. This assumption brings it about that Σ has the same number of dimensions as M. It is further natural to assume that, if the infinitesimal vector $\overrightarrow{OO_1}$ in Σ_P covers $\overrightarrow{PP'}$ on M, then the displacement $\Sigma_P \to \Sigma_{P'}$ will carry O_1 into the center O' of $\Sigma_{P'}$. But no other restrictions should be imposed. Carrying Σ_P by successive steps of infinitesimal displacement around a circuit we shall, when we return to P, have arrived at a definite mapping of Σ_P into itself, an operation of the Klein group that is independent of the choice of coordinates x_i and also, if this is understood in the proper sense, of the choice of admissible frames in all the Klein spaces associated with the various points of M. It will, of course, depend on the circuit described. This automorphism is the generalization of Riemann's curvature. Hence we have here before us the natural general basis on which that notion rests. The infinitesimal trend in geometry initiated by Gauss's theory of curved surfaces now merges with that other line of thought that culminated in Klein's Erlanger program.

It is not advisable to bind the frame of reference in Σ_P to the coordinates x_i covering the neighborhood of P in M. In this respect the old treatment of affinely connected manifolds is misleading. What I said about Cartan's method in dealing with Riemannian geometry was intended as an illustration of this lesson. In studying curves in three-dimensional Euclidean space one does not use a stationary Cartesian frame but associates with the point P traveling along the curve a mobile frame that is adapted to the curve in the most intimate manner, namely the Cartesian frame consisting of tangent, principal normal, and binormal. Freedom means adaptability. Cartan coined the phrase *méthode de répère mobile* for this procedure. Also in the modern development of infinitesimal geometry in the large, where it combines with topology and the associated Klein spaces appear under the name of fibres, it has been found best to keep the *répères*, the frames of the fibre spaces, independent of the coordinates of the underlying manifold.

The temptation is great to mention here some of the endeavors that have been made to utilize these more general geometries for setting up unified field theories encompassing the electromagnetic field beside the gravitational one or even including not only the photons but also the electrons, nucleons, mesons, and whatnot. I shall not succumb to that temptation.

Nobody can predict what sort of geometric structures may be thought up, and hence it would be foolish to claim that our pattern of associated Klein spaces and their displacement is universal. Whatever the structure, it must be described in some arithmetical way relative to a frame of reference $\mathfrak{f}$, whether that frame consists of a coordinate system for the manifold M or of a coordinate system for M plus admissible frames of references for each associated Klein space, or is something even more complicated. Always the problem of equivalence arises, i.e., the question under what conditions two such structures can be carried into each other by a change of the universal frame $\mathfrak{f}$. It was in the attempt of solving this problem for Riemannian geometry that Christoffel first introtuced his 3-indices symbols which later were interpreted by Levi-Civita as describing infinitesimal vector displacement, and that Riemann constructed his curvature tensor. This example is typical. In a number of important cases the attempt of solving the equivalence problem led to associating Klein spaces with the points of the manifold and to defining their infinitesimal displacement. Auxiliary variables that had to be introduced could be interpreted as coordinates in the fibre space Σ_P. Hence in practice the scheme has proved of fairly universal applicability.

I wish to say a word about Cartan's treatment of Riemannian geometry. At each point P of the manifold the vector compass bears a Cartesian frame $\vec{e}_i$. The infinitesimal vector $\overrightarrow{PP'}$ at P is expressed in terms of it,

$$(1) \qquad \overrightarrow{PP'} = dP = \sum \omega_i \vec{e}_i$$

by means of coefficients ω_i that are linear differential forms of the dx_i. Passing from P to the near-by point P' we obtain two Cartesian frames at P': the one that is associated with P' and the one into which the frame associated with P goes over by parallel displacement from P to P'. These two frames in P' are linked by an infinitesimal rotation

$$(2) \qquad \vec{de}_i = \sum_j \omega_{ij} \cdot \vec{e}_j$$

the coefficients ω_{ij} of which are again linear differential forms. The displacement of the vector compass is now described by (1), (2). The condition of closure, which I never formulated, makes the ω_{ij} expressible in terms of the ω_i. The adequate instrument for carrying out the calculations to which such a set-up inevitably leads is a calculus fully developed by Cartan; it deals with multilinear differential forms and their multiplication and "external" derivation. This is a subject of considerable importance in several branches of mathematics including topology.

A scalar $f(x_1, \cdots, x_n)$ has a differential $df = \sum_i \frac{\partial f}{\partial x_i} dx_i$. From an arbitrary linear differential $\varphi = \sum_k \varphi_k dx_k$ we can formally derive $d\varphi = \sum \frac{\partial \varphi_k}{\partial x_i} dx_i dx_k$. The essential point now is that in interpreting this formal expression one assumes the antisymmetric rule $dx_k dx_i = -dx_i dx_k$ for the multiplication of the differentials dx_i. It is then perhaps more sincere to introduce two independent line elements dx and δx and let $d\varphi$ stand for the skew-symmetric bilinear differential form

$$(3) \qquad \sum \left(\frac{\partial \varphi_k}{\partial x_i} - \frac{\partial \varphi_i}{\partial x_k} \right) dx_i \delta x_k.$$

The marvelous thing about this kind of derivative is its invariance with respect to arbitrary coordinate transformations. If φ itself is the differential df of a scalar f, then $d\varphi = 0$: the derivative of a derivative is always zero. Maxwell's theory of the electromagnetic field gives a perfect illustration for this calculus. The potentials φ_i form the coefficients of an invariant differential form $\varphi = \sum \varphi_i dx_i$ of rank 1; its derivative $F = d\varphi$, (3), of rank 2 gives the electromagnetic field strength $F_{ik} = \frac{\partial \varphi_k}{\partial x_i} - \frac{\partial \varphi_i}{\partial x_k}$. Since F itself is a derivative, the form dF of rank 3 must vanish, i.e.,

$$\frac{\partial F_{kl}}{\partial x_i} + \frac{\partial F_{li}}{\partial x_k} + \frac{\partial F_{ik}}{\partial x_l} = 0.$$

The symbolism here employed is not as strange as it may strike you. Indeed look at the customary form of writing a double integral, $\int \cdots dx_1 dx_2$. Here also the proper meaning of $dx_1 dx_2$ as the area of a parallelogram spanned by two line elements $dx, \delta x$ would be more fully exhibited by writing it as the determinant $dx_1 \delta x_2 - dx_2 \delta x_1$; only this form indicates without explanation what happens to the integral under a transformation of coordinates. Multiplication of linear forms

$$\sum f_i dx_i \cdot \sum g_k dx_k = \sum f_i g_k dx_i dx_k$$

must of course also be submitted to the antisymmetric rule $dx_k dx_i = -dx_i dx_k$. It is clear how to pass on to higher ranks. All the integral theorems of vector and tensor analysis are special cases of the general Stokes formula which deals with the integral of a differential form of rank r over an r-dimensional (orientable) manifold imbedded in the space with the coordinates x_i: the integral of the derivative df over a manifold C of dimensionality $r+1$ equals the integral of f over the r-dimensional boundary ∂C of C. The law that the derivative of a derivative vanishes is thus the dual counterpart of the topological law that the boundary ∂C of something is always closed, i.e., has the boundary zero. The question whether a form f, the derivative of which vanishes, is itself a derivative leads straight to the topological theory of homologies and cohomologies; de Rham's work is fundamental in that respect.

An impressive example demonstrating the power of this technique in which the *méthode de répère mobile* combines with the calculus of linear differential forms is a brief paper by Chern [1945], in which he gives an intrinsic proof for the analogue of the Gauss-Bonnet formula in a Riemannian space of arbitrary even dimension. The classical Gauss-Bonnet formula states for a closed surface in three-dimensional Euclidean space that the integral of its Gaussian curvature equals $2\pi \cdot q$ where the even integer q is the most important topological invariant, the Euler characteristic. Allendoerfer had derived the formula for a Riemannian space of even dimension imbedded in Euclidean space, Chern freed it from the imbedding space.[7] It is this perhaps the simplest instance of a relation between the differential and the topological properties of a space, and it seems that there are still many deep problems to solve in this field.

If one tries to understand what is behind the formal apparatus of tensor calculus that is used in general relativity, one arrives with necessity at the general notion of a *covariant quantity*. Let us take those transformations of the Klein space Σ mediating between admissible frames of reference which leave the center O of Σ fixed. They form a group Γ. A covariant quantity of definite type $\mathfrak{C}$ is described relative to an admissible frame $\mathfrak{f}$ by a number of components $X_1, \cdots, X_h$ which vary independently over all real values while the quantity ranges over the manifold of all its possible values. The components X_i' of the same quantity relative to another admissible frame $\mathfrak{f}'$ are connected with the X_i by a linear transformation $X_i' = \sum_j t_{ij} X_j$ the coefficients t_{ij} of which are determined by the operation S of the group Γ that carries $\mathfrak{f}$ into $\mathfrak{f}'$. Composition of the group elements S must be reflected in the composition of the corresponding linear transformations $\|t_{ij}(S)\|$. We then speak of a representation of the group Γ by linear substitutions, and that representation defines the type $\mathfrak{C}$ of the covariant quantity. In the last decades a quite elaborate theory of representations of continuous Lie groups has developed, in which algebraic, differential, and integral methods are blended with each other in a fascinating manner. Here those problems which according to Study's complaint the relativists had let go by the board are attacked on a much deeper level than the formalistically minded Study had ever dreamt of. For the representations of the linear group the symmetry characters studied in A. Young's Quantitative Analysis and in a different manner by G. Frobenius, proved of great importance. Their bearing upon the quantum mechanics of systems consisting of equal particles (e.g., electrons) has been disclosed by Wigner. For the orthogonal group Cartan found a host of double-valued irreducible representations not less numerous than the single-valued ones. Their appearance is due to the topological fact that the orthogonal group is not simply connected but has a simply connected covering manifold of two sheets extending over it without boundaries and ramifications. The most elementary of these double-valued representations is the spin representation which Dirac used in his Lorentz-invariant quantum theory of the electron.

Of course it would be foolish to maintain that all these investigations have their origin in relativity theory. Indeed Frobenius and Issai Schur's spadework on finite and compact groups and Cartan's early work on semi-simple Lie groups and their representations had nothing to do with it. But for myself I can say that the wish to understand what really is the mathematical substance behind the formal apparatus of relativity theory led me to the study of representations and invariants of groups; and my experience in this regard is probably not unique.

What is the upshot of it all? Relativity theory is intimately intertwined with a number of important branches of mathematics. Its influence in mathematics has been far less revolutionary than in physics and the epistemology of natural science; for its pattern fitted perfectly into the pattern of ideas already current in mathematics. But just because it could be absorbed so readily by mathematics it has stimulated the development and elaboration of those mathematical ideas to which it had a natural affinity.

Notes

[Weyl prepared this lecture for a celebration at the Institute for Advanced Study marking the seventieth birthday of Einstein, March 19, 1949; it was published in Weyl 1949b and *WGA* 4:394–400.]

1 [Regarding the Erlangen program (which Weyl here calls Erlanger), see above 66.]

2 [In 1949, many physicists considered general relativity a purely theoretical subject, important for cosmology, but not having any feasible experimental tests. That began to change with the experiments of R. V. Pound, G. A. Rebka, and J. L. Snider (Pound and Rebka 1959), which demonstrated the gravitational red-shift in a terrestrial experiment using the Mössbauer effect.]

3 [In contrast to Einstein's well-known resistance to the quantum theory, Weyl was very open to it and continued to work on applying his gauge field ideas to Dirac spinors. Weyl 1929a,b introduced the tetrad or vierbein formulation of general relativity, meant to facilitate its connection with the Dirac theory.]

4 [Weyl 1952a, 112–117, 124–125, discusses affine manifolds in detail.]

5 [Weyl discusses the implications of conformal invariance for the dimensionality of space-time below, 212–213.]

6 [Weyl 1929c discusses the relation of his work and Cartan's, continued in Brauer and Weyl 1935. Einstein was very interested in Cartan and corresponded with him on mathematical extensions of general relativity; see Cartan 1931, 1983, 1986, Pesic 2007, 179–186, and Cartan and Einstein 1979. For Weyl's relation to Cartan, see Scholz 2010.]

7 As a matter of fact, before him Allendoerfer and A. Weil had given a proof by embedding each of the cells into which the space is cut up into a Euclidean space. Chern got rid of this embedding device.

A Half-Century of Mathematics

(1951)

1. Introduction. Axiomatics.

Mathematics, beside astronomy, is the oldest of all sciences. Without the concepts, methods and results found and developed by previous generations right down to Greek antiquity, one cannot understand either the aims or the achievements of mathematics in the last fifty years. Mathematics has been called the science of the infinite; indeed, the mathematician invents finite constructions by which questions are decided that by their very nature refer to the infinite. That is his glory. Kierkegaard once said religion deals with what concerns man unconditionally. In contrast (but with equal exaggeration) one may say that mathematics talks about the things which are of no concern at all to man. Mathematics has the inhuman quality of starlight, brilliant and sharp, but cold. But it seems an irony of creation that man's mind knows how to handle things the better the farther removed they are from the center of his existence. Thus we are cleverest where knowledge matters least: in mathematics, especially in number theory. There is nothing in any other science that, in subtlety and complexity, could compare even remotely with such mathematical theories as for instance that of algebraic class fields. Whereas physics in its development since the turn of the century resembles a mighty stream rushing on in one direction, mathematics is more like the Nile delta, its waters fanning out in all directions. In view of all this: dependence on a long past, other-worldliness, intricacy, and diversity, it seems an almost hopeless task to give a non-esoteric account of what mathematicians have done during the last fifty years. What I shall try to do here is, first to describe in somewhat vague terms general trends of development, and then in more precise language explain the most outstanding mathematical notions devised, and list some of the more important problems solved, in this period.

One very conspicuous aspect of twentieth-century mathematics is the enormously increased role which the axiomatic approach plays. Whereas the axiomatic method was formerly used merely for the purpose of elucidating the foundations on which we build, it has now become a tool for concrete mathematical research. It is perhaps in algebra that it has scored its greatest successes. Take for instance the system of real numbers. It is like a Janus head facing in two

159

directions: on the one side it is the field of the algebraic operations of addition and multiplication; on the other hand it is a continuous manifold, the parts of which are so connected as to defy exact isolation from each other. The one is the algebraic, the other the topological face of numbers. Modern axiomatics, simple-minded as it is (in contrast to modern politics), does not like such ambiguous mixtures of peace and war, and therefore cleanly separated both aspects from each other.

In order to understand a complex mathematical situation it is often convenient to separate in a natural manner the various sides of the subject in question, make each side accessible by a relatively narrow and easily surveyable group of notions and of facts formulated in terms of these notions, and finally return to the whole by uniting the partial results in their proper specialization. The last synthetic act is purely mechanical. The art lies in the first, the analytic act of suitable separation and generalization. Our mathematics of the last decades has wallowed in generalizations and formalizations. But one misunderstands this tendency if one thinks that generality was sought merely for generality's sake. The real aim is simplicity: every natural generalization simplifies since it reduces the assumptions that have to be taken into account. It is not easy to say what constitutes a natural separation and generalization. For this there is ultimately no other criterion but fruitfulness: the success decides. In following this procedure the individual investigator is guided by more or less obvious analogies and by an instinctive discernment of the essential acquired through accumulated previous research experience. When systematized the procedure leads straight to axiomatics. Then the basic notions and facts of which we spoke are changed into undefined terms and into axioms involving them. The body of statements deduced from these hypothetical axioms is at our disposal now, not only for the instance from which the notions and axioms were abstracted, but wherever we come across an interpretation of the basic terms which turns the axioms into true statements. It is a common occurrence that there are several such interpretations with widely different subject matter.

The axiomatic approach has often revealed inner relations between, and has made for unification of methods within, domains that apparently lie far apart. This tendency of several branches of mathematics to coalesce is another conspicuous feature in the modern development of our science, and one that goes side by side with the apparently opposite tendency of axiomatization. It is as if you took a man out of a milieu in which he had lived not because it fitted him but from ingrained habits and prejudices, and then allowed him, after thus setting him free, to form associations in better accordance with his true inner nature.

In stressing the importance of the axiomatic method I do not wish to exaggerate. Without inventing new constructive processes no mathematician will get very far. It is perhaps proper to say that the strength of modern mathematics lies in the interaction between axiomatics and construction. Take algebra as a representative example. It is only in this century that algebra has come into its own by breaking away from the one universal system Ω of numbers which used to form the basis of all mathematical operations as well as all physical measurements. In its newly-acquired freedom algebra envisages an infinite variety of "number fields" each of which may serve as an operational basis; no attempt is made to embed them into the one system Ω. Axioms limit the possibilities for the number concept; constructive processes yield number fields that satisfy the axioms.

In this way algebra has made itself independent of its former master analysis and in some branches has even assumed the dominant role. This development in mathematics is paralleled in physics to a certain degree by the transition from classical to quantum physics, inasmuch as the latter ascribes to each physical structure its own system of observables or quantities. These quantities are subject to the algebraic operations of addition and multiplication; but as their multiplication is non-commutative, they are certainly not reducible to ordinary numbers.

At the International Mathematical Congress in Paris in 1900 David Hilbert, convinced that problems are the life-blood of science, formulated twenty-three unsolved problems which he expected to play an important role in the development of mathematics during the next era. How much better he predicted the future of mathematics than any politician foresaw the gifts of war and terror that the new century was about to lavish upon mankind! We mathematicians have often measured our progress by checking which of Hilbert's questions had been settled in the meantime. It would be tempting to use his list as a guide for a survey like the one attempted here.[1] I have not done so because it would necessitate explanation of too many details. I shall have to tax the reader's patience enough anyhow.

Part I. Algebra. Number Theory. Groups.

2. Rings, Fields, Ideals.

Indeed, at this point it seems impossible for me to go on without illustrating the axiomatic approach by some of the simplest algebraic notions. Some of them are as old as Methuselah. For what is older than the sequence of *natural numbers* 1, 2, 3, $\ldots$, by which we count? Two such numbers a, b may be added and multiplied ($a + b$ and $a \cdot b$). The next step in the genesis of numbers adds to these positive *integers* the negative ones and zero; in the wider system thus created the operation of addition permits of a unique inversion, subtraction. One does not stop here: the integers in their turn get absorbed into the still wider range of *rational numbers* (fractions). Thereby division, the operation inverse to multiplication, also becomes possible, with one notable exception however: division by zero. (Since $b \cdot 0 = 0$ for every rational number b, there is no inverse b of 0 such that $b \cdot 0 = 1$.) I now formulate the fundamental facts about the operations "plus" and "times" in the form of a table of axioms:

TABLE T

---

(1) The cumulative and associative laws for addition,

$$a + b = b + a \qquad\qquad a + (b + c) = (a + b) + c$$

(2) The corresponding laws for multiplication.

(3) The distributive law connecting addition with multiplication,

$$c \cdot (a + b) = (c \cdot a) + (c \cdot b)$$

(4) The axioms of subtraction: (4_1) There is an element o (0, "zero") such that $a + o = o + a = a$ for every a. (4_2) To every a there is a number $-a$ such that $a + (-a) = (-a) + a = o$.

(5) The axioms of division: (5_1) There is an element e (1, "unity") such that $a \cdot e = e \cdot a = a$ for every a. (5_2) To every $a \neq o$ there is an a^{-1} such that $a \cdot a^{-1} = a^{-1} \cdot a = e$.

By means of (4_2) and (5_2) one may introduce the difference $b - a$ and the quotient b/a as $b + (-a)$ and $b \cdot a^{-1}$, respectively.

When the Greeks discovered that the ratio ($\sqrt{2}$) between diagonal and side of a square is not measurable by a rational number, a further extension of the number concept was called for. However, all measurements of continuous quantities are possible only approximately, and always have a certain range of inaccuracy. Hence rational numbers, or even finite decimal fractions, can and do serve the ends of mensuration provided they are interpreted as approximations, and a calculus with approximate numbers seems the adequate numerical instrument for all measuring sciences. But mathematics ought to be prepared for any subsequent refinement of measurements. Hence dealing, say, with electric phenomena, one would be glad if one could consider the approximate values of the charge e of the electron which the experimentalist determines with ever greater accuracy as approximations of one definite *exact* value e. And thus, during more than two millenniums from Plato's time until the end of the nineteenth century, the mathematicians worked out an exact number concept, that of *real numbers*, that underlies all our theories in natural science. Not even to this day are the logical issues involved in that concept completely clarified and settled. The rational numbers are but a small part of the real numbers. The latter satisfy our axioms no less than the rational ones, but their system possesses a certain completeness not enjoyed by the rational numbers, and it is this, their "topological" feature, on which the operations with infinite sums and the like, as well as all continuity arguments, rest. We shall come back to this later.

Finally, during the Renaissance *complex numbers* were introduced. They are essentially pairs $z = (x, y)$ of real numbers x, y, pairs for which addition and multiplication are defined in such a way that all axioms hold. On the ground of these definitions $e = (1, 0)$ turns out to be the unity, while $i = (0, 1)$ satisfies the equation $i \cdot i = -e$. The two members x, y of the pair z are called its real and imaginary parts, and z is usually written in the form $xe + yi$, or simply $x + yi$. The usefulness of the complex numbers rests on the fact that every algebraic equation (with real or even complex coefficients) is solvable in the field of complex numbers. The analytic functions of a complex variable are the subject of a particularly rich and harmonious theory, which is the show-piece of classical nineteenth century analysis.

A set of elements for which the operations $a + b$ and $a \cdot b$ are so defined as to satisfy the axioms (1)–(4) is called a *ring*; it is called a *field* if also the axioms (5) hold. Thus the common integers form a ring I, the rational numbers form a field ω; so do the real numbers (field Ω) and the complex numbers (field Ω^*). But these are by no means the only rings or fields. The polynomials of all possible degrees h,

$$(1) \qquad f = f(x) = a_0 + a_1 x + a_2 x^2 + \cdots + a_h x^h,$$

with coefficients a_i taken from a given ring R (e.g. the ring I of integers, or the field ω), called "polynomials over R," form a ring $R[x]$. Here the variable or indeterminate x is to be looked upon as an empty symbol; the polynomial is really nothing but the sequence of its coefficients $a_0, a_1, a_2, \ldots$. But writing it in the customary form (1) suggests the rules for the addition and multiplication of polynomials which I will not repeat here. By substituting for the variable x a definite element ("number") y of R, or of a ring P containing R as a subring, one projects the elements f of $R[x]$ into elements α of $P, f \to \alpha$: the polynomial $f = f(x)$ goes over into the number $\alpha = f(\gamma)$. This mapping $f \to \alpha$ is *homomorphic*, i.e., it preserves addition and multiplication. Indeed, if the substitution of γ for x carries the polynomial f into α and the polynomial g into β then it carries $f + g, f \cdot g$ into $\alpha + \beta, \alpha \cdot \beta$, respectively.

If the product of two elements of a ring is never zero unless one of the factors is, one says that the ring is without null-divisor. This is the case for the rings discussed so far. A field is always a ring without null-divisor. The construction by which one rises from the integers to the fractions can be used to show that any ring R with unity and without null-divisor may be imbedded in a field k, the quotient field, such that every element of k is the quotient a/b of two elements a and b of R, the second of which (the denominator) is not zero.

Writing $1a, 2a, 3a, \cdots$, for $a, a+a, a+a+a$, etc., we use the natural numbers $n = 1, 2, 3, \cdots$, as multipliers for the elements a of a ring or a field. Suppose the ring contains the unity e. It may happen that a certain multiple ne of e equals zero; then one readily sees that $na = 0$ for every element a of the ring. If the ring is without null divisors, in particular if it is a field and p is the least natural number for which $pe = 0$, then p is necessarily a prime number like 2 or 3 or 5 or 7 or 11 $\cdots$. One thus distinguishes fields of prime characteristic p from those of characteristic 0 in which no multiple of e is zero.

Plot the integers $\cdots, -2, -1, 0, 1, 2, \ldots$ as equidistant marks on a line. Let n be a natural number ≥ 2 and roll this line upon a wheel of circumference n. Then any two marks a, a' coincide, the difference $a - a'$ of which is divisible by n. (The mathematicians write $a = a' \pmod{n}$; they say: a congruent to a' modulo n.) By this identification the ring of integers I goes over into a ring I_n, consisting of n elements only (the marks on the wheel), as which one may take the "residues" $0, 1, \cdots, n - 1$. Indeed, congruent numbers give congruent results under both addition and multiplication: $a = a', b = b' \pmod{n}$ imply $a + b = a' + b', a \cdot b = a' \cdot b' \pmod{n}$. For instance, modulo 12 we have $7 + 8 = 3, 5 \cdot 8 = 4$ because 15 leaves the residue 3 and 40 the residue 4 if divided by 12. The ring I_{12} is not without null divisors since $3 \cdot 4$ is divisible by 12, but neither 3 nor 4 is. However, if p is a natural prime number, then I_p has no null divisor and is even a field; for as the ancient Greeks proved by an ingenious procedure (Euclid's algorism), for every integer a not divisible by p there is one, a', such that $a \cdot a' = 1$ $\pmod{p}$. This Euclidean theorem is at the basis of the whole of number theory. The example shows that there are fields of any given prime characteristic p.

In any ring R one may introduce the notions of unit and prime element as follows. The ring element a is a unit if it has a reciprocal a' in the ring, such that $a' \cdot a = e$. The element a is composite if it may be decomposed into two factors $a_1 a_2$, neither of which is a unit. A prime number is one that is neither a unit nor composite. The units of I are the numbers $+1$ and -1. The units of the ring $k[x]$ of polynomials over a field k are the non-vanishing elements of k (polynomials of degree 0). According to the Greek discovery of the irrationality of $\sqrt{2}$ the polynomial $x^2 - 2$ is prime in the ring $\omega[x]$; but, of course, not in $\Omega[x]$, for there it splits into the two linear factors $(x - \sqrt{2})(x + \sqrt{2})$. Euclid's algorism is also applicable to polynomials $f(x)$ of one variable x over any field k. Hence they satisfy Euclid's theorem: Given a prime element $P = P(x)$ in this ring $k[x]$ and an element $f(x)$ of $k[x]$ not divisible by $P(x)$, there exists another polynomial $f'(x)$ over k such that $\{f(x) \cdot f'(x)\} - 1$ is divisible by $P(x)$. Identification of any elements f and g of $k[x]$, the difference of which is divisible by P, therefore changes the ring $k[x]$ into a field, the "residue field κ of $k[x]$ modulo P." Example: $\omega[x] \bmod x^2 - 2$. (Incidentally the complex numbers may be described as the elements of the residue field of $\Omega[x] \bmod x^2 + 1$.) Strangely enough, the fundamental Euclidean theorem does not hold for polynomials of two variables x, y. For instance, $P(x, y) = x - y$ is a prime element of $\omega[x, y]$, and $f(x, y) = x$ an element not divisible by $P(x, y)$. But a congruence

$$x \cdot f'(x, y) \equiv 1 \pmod{x - y}$$

is impossible. Indeed, it would imply $-1 + x \cdot f'(x, x) = 0$, contrary to the fact that the polynomial of one indeterminate x,

$$-1 + x \cdot f'(x, x) = -1 + c_1 x + c_2 x^2 + \cdots,$$

is not zero. Thus the ring $\omega[x, y]$ does not obey the simple laws prevailing in I and in $\omega[x]$.

Consider κ, the residue field of $\omega[x]$ mod $x^2 - 2$. Since for any two polynomials $f(x), f'(x)$ which are congruent mod $x^2 - 2$ the numbers $f(\sqrt{2}), f'(\sqrt{2})$ coincide, the transition $f(x) \to f(\sqrt{2})$ maps κ into a sub-field $\omega[\sqrt{2}]$ of Ω consisting of the numbers $a + b\sqrt{2}$ with rational a, b. Another such projection would be $f(x) \to f(-\sqrt{2})$. In former times one looked upon κ as the part $\omega[\sqrt{2}]$ of the continuum Ω or Ω^* of all real or all complex numbers; one wished to embed everything into this universe Ω or Ω^* in which analysis and physics operate. But as we have introduced it here, κ is an algebraic entity the elements of which are not numbers in the ordinary sense. It requires for its construction no other numbers but the rational ones. It has nothing to do with Ω, and ought not to be confused with the one or the other of its two projections into Ω. More generally, if $P = P(x)$ is any prime element in $\omega[x]$ we can form the residue field κ_P of $\omega[x]$ modulo P. To be sure, if δ is any of the real or complex roots of the equation $P(x) = 0$ in Ω^* then $f(x) \to f(\delta)$ defines a homomorphic projection of κ_P into Ω^*. But the projection is not κ_P itself.

Let us return to the ordinary integers $\cdots, -2, -1, 0, 1, 2, \cdots$, which form the ring I. The multiples of 5, i.e., the integers divisible by 5, clearly form a ring. It is a ring without unity, but it has another important peculiarity: not only does the product of any two of its elements lie in it, but all the integral multiples of an element do. The queer term *ideal* has been introduced for such a set: Given a ring R, an R-ideal $(\mathfrak{a})$ is a set of elements of R such that (1) sum and difference of any two elements of $(\mathfrak{a})$ are in $(\mathfrak{a})$, (2) the product of an element in $(\mathfrak{a})$ by any element of R is in $(\mathfrak{a})$: We may try to describe a divisor $\mathfrak{a}$ by the set of all elements divisible by $\mathfrak{a}$. One would certainly expect this set to be an ideal $(\mathfrak{a})$ in the sense just defined. Given an ideal $(\mathfrak{a})$, there may not exist an actual element a of R such that $(\mathfrak{a})$ consists of all multiples $j = m \cdot a$ of a (m any element in R). But then we would say that $(\mathfrak{a})$ stands for an "ideal divisor" $\mathfrak{a}$: the words "the element j of R is divisible by $\mathfrak{a}$" would simply mean: "j belongs to $(\mathfrak{a})$." In the ring I of common integers all divisors are actual.

But this is not so in every ring. An algebraic surface in the three-dimensional Euclidean space with the Cartesian coordinates x, y, z is defined by an equation $F(x, y, z) = 0$ where F is an element of $^3\Omega = \Omega[x, y, z]$, i.e., a polynomial of the variables x, y, z with real coefficients. F is zero in all the points of the surface; but the same is true for every multiple $L \cdot F$ of F (L being any element of $^3\Omega$), in other words, for every polynomial of the ideal (F) in $^3\Omega$. Two simultaneous polynomial equations

$$F_1(x, y, z) = 0, \qquad F_2(x, y, z) = 0$$

will in general define a curve, the intersection of the surface $F_1 = 0$ and the surface $F_2 = 0$. The polynomials $(L_1 \cdot F_l) + (L_2 \cdot F_2)$ formed by arbitrary elements L_1, L_2 of $^3\Omega$ form an ideal (F_1, F_2), and all these polynomials vanish on the curve. This ideal will in general not correspond to an actual divisor F, for a curve is not a surface. Examples like this should convince the reader that the study of algebraic manifolds (curves, surfaces, etc., in 2, 3, or any number of dimensions) amounts essentially to a study of polynomial ideals. The field of coefficients is not necessarily Ω or Ω^*, but may be a field of a more general nature.

3. Some achievements of algebra and number theory.

I have finally reached a point where I can hint, I hope, with something less than complete obscurity, at some of the accomplishments of algebra and number theory in our century. The most important is probably the freedom with which we have learned to manage these abstract axiomatic concepts, like field, ring, ideal, etc. The atmosphere in a book like van der Waerden's *Moderne Algebra*, published about 1930 [van der Waerden 1970], is completely different from that prevailing, e.g., in the articles on algebra written for the *Mathematical Encyclopædia* around 1900. More specifically, a general theory of ideals, and in particular of polynomial ideals, was developed. (However, it should be said that the great pioneer of abstract algebra, Richard Dedekind, who first introduced the ideals into number theory, still belonged to the nineteenth century.) Algebraic geometry, before and around 1900 flourishing chiefly in Italy, was at that time a discipline of a type uncommon in the sisterhood of mathematical disciplines: it had powerful methods, plenty of general results, but they were of somewhat doubtful validity. By the abstract algebraic methods of the twentieth century all this was put on a safe basis, and the whole subject received a new impetus. Admission of fields other than Ω^*, as the field of coefficients, opened up a new horizon.

A new technique, the "primadic numbers," was introduced into algebra and number theory by K. Hensel shortly after the turn of the century, and since then has become of ever increasing importance.[2] Hensel shaped this instrument in analogy to the power series which played such an important part in Riemann's and Weierstrass's theory of algebraic functions of one variable and their integrals (Abelian integrals). In this theory, one of the most impressive accomplishments of the previous century, the coefficients were supposed to vary over the field Ω^* of all complex numbers. Without pursuing the analogy, I may illustrate the idea of p-adic numbers by one typical example, that of quadratic norms. Let p be a prime number, and let us first agree that a congruence $a = b$ modulo a power p^h of p for rational numbers a, b has this meaning that $(a - b)/p^h$ equals a fraction whose denominator is not divisible by p;

$$\text{e.g.,} \frac{39}{4} - \frac{12}{5} \equiv 0 \pmod{7^2} \quad \text{because} \quad \frac{39}{4} - \frac{12}{5} = 7^2 \cdot \frac{3}{20}.$$

Let now a, b be rational numbers, $a \neq 0$, and b not the square of a rational number. In the quadratic field $\omega[\sqrt{b}]$ the number a is a *norm* if there are rational numbers x, y such that

$$a = (x + y\sqrt{b})(x - y\sqrt{b}), \text{ or } a = x^2 - by^2.$$

Necessary for the solvability of this equation is (1) that for every prime p and every power p^h of p the congruence $a \equiv x^2 - by^2 \pmod{p^h}$ has a solution. This is what we mean by saying the equation has a p-adic solution. Moreover there must exist rational numbers x and y such that $x^2 - by^2$ differs as little as one wants from a. This is what we mean by saying that the equation has an ∞-adic solution. The latter condition is clearly satisfied for every a provided b is positive; however, if b is negative it is satisfied only for positive a. In the first case every a is ∞-adic norm, in the second case only half of the a's are, namely, the positive ones. A similar situation prevails with respect to p-adic norms. One proves that these necessary conditions are also sufficient: if a is a norm locally everywhere, i.e., if $a = x^2 - by^2$ has a p-adic solution for every "finite prime spot p" and also for the "infinite prime spot ∞," then it has a "global" solution, namely an exact solution in rational numbers x, y.

This example, the simplest I could think of, is closely connected with the theory of general of quadratic forms, a subject that goes back to Gauss' *Disquisitiones arithmeticae*, but in which the twentieth century has made some decisive progress by means of the p-adic technique, and it is also typical for that most fascinating branch of mathematics mentioned in the introduction: class field theory. Around 1900 David Hilbert had formulated a number of interlaced theorems concerning class fields, proved some of them at least in special cases, and left the rest to his twentieth century successors, among whom I name Takagi, Artin and Chevalley. His norm residue symbol paved the way for Artin's general reciprocity law. Hilbert had used the analogy with the Riemann-Weierstrass theory of algebraic functions over Ω^* for his orientation, but the ingenious, partly transcendental methods which he applied had nothing to do with the much simpler ones that had proved effective for the functions. By the primadic technique a rapprochement of methods has occurred, although there is still a considerable gap separating the theory of algebraic functions and the much subtler algebraic numbers.

Hensel and his successors have expressed the p-adic technique in terms of the non-algebraic "topological" notion of ("valuation" or) *convergence*. An infinite sequence of rational numbers $a_1, a_2, \cdots$ is convergent if the difference $a_i - a_j$ tends to zero, $a_i - a_j \to 0$, provided i and j independently of each other tend to infinity; more explicitly, if for every positive rational number ϵ there exists a positive integral N such that $-\epsilon < a_i - a_j < \epsilon$ for all i and $j > N$. The completeness of the real number system is expressed by Cauchy's convergence theorem: To every convergent sequence $a_1, a_2, \cdots$ of rational numbers there exists a *real* number α to which it converges: $a_i - \alpha \to 0$ for $i \to \infty$. With the ∞-adic concept of convergence we have now confronted the p-adic one induced by a prime number p. Here the sequence is considered convergent if for every exponent $h = 1, 2, 3, \cdots$, there is a positive integer N such that $a_i - a_j$ is divisible by p^h as soon as i and $j > N$. By introduction of p-adic numbers one can make the system of rational numbers complete in the p-adic sense as the introduction of real numbers makes them complete in the ∞-adic sense. The rational numbers are embedded in the continuum of all real numbers, but they may be embedded as well in that of all p-adic numbers. Each of these embedments corresponding to a finite or the infinite prime spot p is equally interesting from the arithmetical viewpoint. Now it is more evident than ever how wrong it was to identify an algebraic number field with one of its homomorphic projections into the field Ω of real numbers; along with the (real) infinite prime spots one must pay attention to the finite prime spots which correspond to the various prime ideals of the field. This is a golden rule abstracted from earlier, and then made fruitful for later, arithmetical research; and here is one bridge (others will be pointed out later) joining the two most fascinating branches of modern mathematics: abstract algebra and topology.

Besides the introduction of the primadic treatment and the progress made in the theory of class fields, the most important advances of number theory during the last fifty years seem to lie in those regions where the powerful tool of analytic functions can be brought to bear upon its problems. I mention two such fields of investigation: I. distribution of primes and the zeta function, II. additive number theory.

I. The notion of prime number is of course as old and as primitive as that of the multiplication of natural numbers. Hence it is most surprising to find the distribution of primes among all natural numbers is of such a highly irregular and almost mysterious character. While on the whole the prime numbers thin out the further one gets in the sequence of numbers, wide gaps are always followed again by clusters. An old conjecture of Goldbach's maintains that there even come along again and again pairs of primes of the smallest possible difference 2, like 57 and 59. However, the distribution of primes obeys at least a fairly simple *asymptotic* law: the number $\pi(n)$ of primes among all numbers from 1 to n is asymptotically equal to $n/\log n$. (Here $\log n$ is not the Briggs logarithm which our logarithmic tables give, but the natural logarithm as defined by the integral $\int_1^n dx/x$.) By asymptotic is meant that the quotient between $\pi(n)$ and the approximating function $n/\log n$ tends to 1 as n tends to infinity. In antiquity Eratosthenes had devised a method to sift out the prime numbers. By this sieve method the Russian mathematician Tchebycheff had obtained, during the nineteenth century, the first non-trivial results about the distribution of primes. Riemann used a different approach: his tool is the so-called zeta-function defined by the infinite series

(2) $$\zeta(s) = 1^{-s} + 2^{-s} + 3^{-s} + \cdots .$$

Here s is a complex variable, and the series converges for all values of s, the real part of which is greater than 1, $\Re s > 1$. Already in the eighteenth century the fact that every positive integer can be uniquely factorized into primes had been translated by Euler into the equation

$$1/\zeta(s) = (1 - 2^{-s})(1 - 3^{-s})(1 - 5^{-s}) \cdots$$

where the (infinite) product extends over all primes $2, 3, 5, \cdots$. Riemann showed that the zeta-function has a unique "analytic continuation" to all values of s and that it satisfies a certain functional equation connecting its values for s and $1 - s$. Decisive for the prime number problem are the zeros of the zeta-function, i.e., the values s for which $\zeta(s) = 0$. Riemann's equation showed that, except for the "trivial" zeros at $s = -2, -4, -6, \cdots$, all zeros have real parts between 0 and 1. Riemann conjectured that their real parts actually equal $\frac{1}{2}$. His conjecture has remained a challenge to mathematics now for almost a hundred years. However, enough had been learned about these zeros at the close of the nineteenth century to enable mathematicians, by means of some profound and newly-discovered theorems concerning analytic functions, to prove the above-mentioned asymptotic law. This was generally considered a great triumph of mathematics. Since the turn of the century Riemann's functional equation with the attending consequences has been carried over from the "classical" zeta-function (ii) of the field of rational numbers to that of an arbitrary algebraic number field (E. Hecke). For certain fields of prime characteristic one succeeded in confirming Riemann's conjecture, but this provides hardly a clue for the classical case. About the classical zeta-function we know now that it has infinitely many zeros on the critical line $\Re s = \frac{1}{2}$, and even that at least a fixed percentage, say 10 per cent, of them lie on it. (What this means is the following: Some percentage of those zeros whose imaginary part lies between arbitrary fixed limits $-T$ and $+T$ will have a real part equal to $\frac{1}{2}$, and this percentage will not sink below a certain positive limit, like 10 per cent, when T tends to infinity.) Finally about two years ago Atle Selberg succeeded, to the astonishment of the mathematical world, in giving an "elementary" proof of the prime number law by an ingenious refinement of old Eratosthenes' sieve method.[3]

II. It has been known for a long time that every natural number n may be written as the sum of at most four square numbers, e.g.,

$$7 = 2^2 + 1^2 + 1^2 + 1^2, \quad 87 = 9^2 + 2^2 + 1^2 + 1^2 = 7^2 + 5^2 + 3^2 + 2^2.$$

The same question arises for cubes, and generally for any kth powers ($k = 2, 3, 4, 5, \ldots$). In the eighteenth century Waring had conjectured that every nonnegative integer n may be expressed as the sum of a limited number M of kth powers,

$$(3) \qquad n = n_1^k + n_2^k + \cdots + n_M^k,$$

where the n_i are also non-negative integers and M is independent of n. The first decade of the twentieth century brought two events: first one found that every n is expressible as the sum of at most 9 cubes (and that, excepting a few comparatively small n, even 8 cubes will do); and shortly afterwards Hilbert proved Waring's general theorem. His method was soon replaced by a different approach, the Hardy-Littlewood circle method, which rests on the use of a certain analytic function of a complex variable and yields asymptotic formulas for the number of different representations of n in the form (3). With some precautions demanded by the nature of the problem, and by overcoming some quite serious obstacles, the result was later carried over to arbitrary algebraic number fields; and by a further refinement of the circle method in a different direction Vinogradoff proved that every sufficiently large n is the sum of at most 3 primes. Is it even true that every even n is the sum of 2 primes? To show this seems to transcend our present mathematical powers as much as Goldbach's conjecture. The prime numbers remain very elusive fellows.

III. Finally, a word ought to be said about investigations concerning the arithmetical nature of numbers originating in analysis. One of the most elementary such constants is π, the area of the circle of radius 1. By proving that π is a transcendental number (not satisfying an algebraic equation with rational coefficients) the age-old problem of "squaring the circle" was settled in 1882 in the negative sense; that is, one cannot square the circle by constructions with ruler and compass. In general it is much harder to establish the transcendency of numbers than of functions. Whereas it is easy to see that the exponential function

$$e^x = 1 + \frac{x}{1} + \frac{x^2}{1 \cdot 2} + \frac{x^3}{1 \cdot 2 \cdot 3} + \cdots$$

is not algebraic, it is quite difficult to prove that its basis e is a transcendental number. C. L. Siegel was the first who succeeded, around 1930, in developing a sort of general method for testing the transcendency of numbers. But the results in this field remain sporadic.[4]

4. Groups, vector spaces and algebras.

This ends our report on number theory, but not on algebra. For now we have to introduce the *group* concept, which, since the young genius Évariste Galois blazed the trail in 1830, has penetrated the entire body of mathematics. Without it an understanding of modern mathematics is impossible. Groups first occurred as *groups of transformations*. Transformations may operate in any set of elements, whether it is finite like the integers from 1 to 10, or infinite like the points in space. *Set* is a premathematical concept: whenever we deal with a realm of objects, a set is defined by giving a criterion which decides for any object of the realm whether it belongs to the set or not. Thus we speak of the set of prime numbers, or of the set of all points on a circle, or of all points with rational coordinates in a given coordinate system, or of all people living at this moment in the State of New Jersey. Two sets are considered equal if every element of the one belongs to the other and vice versa. A *mapping S* of a set σ into a set σ' is defined if with every element a of σ there is associated an element a' of σ', $a \to a'$. Here a rule is required which allows one to find the "image" a' for any given element a of σ. This general notion of mapping we may also call of a premathematical nature. Examples: a real-valued function of a real variable is a mapping of the continuum Ω into itself. Perpendicular projection of the space points upon a given plane is a mapping of the space into the plane. Representing every space-point by its three coordinates x, y, z with respect to a given coordinate system is a mapping of space into the continuum of real number triples (x, y, z). If a mapping $S, a \to a'$ of σ into σ', is followed by a mapping $S', a' \to a''$ of σ' into a third set σ'', the result is a mapping $SS' : a \to a''$ of σ into σ''. A *one-to-one mapping* between two sets σ, σ' is a pair of mappings, $S : a \to a'$ of σ into σ', and $S' : a' \to a$ of σ' into σ, which are inverse to each other. This means that the mapping SS' of σ into σ is the identical mapping E of σ which sends every element a of σ into itself, and that $S'S$ is the identical mapping of σ'. In particular, one is interested in one-to-one mappings of a set σ into itself. For them we shall use the word *transformation*. Permutations are nothing but transformations of a finite set.

The inverse S' of a transformation $S, a \to a'$ of a given set σ, is again a transformation and is usually denoted by S^{-1}. The result ST of any two transformations S and T of σ is again a transformation, and its inverse is $T^{-1}S^{-1}$ (according to the rule of dressing and undressing: if in dressing one begins with the shirt and ends with the jacket, one must in undressing begin with the jacket and end with the shirt. The order of the two "factors" S, T is essential.) *A group of transformations* is a set of transformations of a given manifold which (1) contains the identity E, (2) contains with every transformation S its inverse S^{-1}, and (3) with any two transformations S, T their "product" ST. Example: One could define congruent configurations in space as point sets of which one goes into the other by a congruent transformation of space. The congruent transformations, or "motions," of space form a group; a statement which, according to the above definition of group, is equivalent to the threefold statement that (1) every figure is congruent to itself, (2) if a figure F is congruent to F', then F' is congruent to F, and (3) if F is congruent to F' and F' congruent to F'', then F is congruent to F''. This example at once illuminates the inner significance of the group concept. *Symmetry* of a configuration F in space is described by the group of motions that carry F into itself.

Often manifolds have a structure. For instance, the elements of a field are connected by the two operations of plus and times; or in Euclidean space we have the relationship of congruence between figures. Hence we have the idea of structure-preserving mappings; they are called *homomorphisms*. Thus a homomorphic mapping of a field k into a field k' is a mapping $a \to a'$ of the "numbers" a of k into the numbers a' of k' such that $(a + b)' = a' + b'$ and $(a \cdot b)' = a' \cdot b'$. A homomorphic mapping of space into itself would be one that carries any two congruent figures into two mutually congruent figures. The following terminology (suggestive to him who knows a little Greek) has been agreed upon: homomorphisms which are one-to-one mappings are called isomorphisms; when a homomorphism maps a manifold σ into itself, it is called an endomorphism, and an automorphism when it is both: a one-to-one mapping of σ into itself. Isomorphic systems, i.e., any two systems mapped isomorphically upon each other, have the same structure; indeed nothing can be said about the structure of the one system that is not equally true for the other.

The *automorphisms* of a manifold with a well-defined structure form a *group*. Two sub-sets of the manifold that go over into each other by an automorphism deserve the name of *equivalent*. This is the precise idea at which Leibniz hints when he says that two such sub-sets are "indiscernible when each is considered in itself"; he recognized this general idea as lying behind the specific geometric notion of similitude. The general problem of relativity consists in nothing else but to find the group of automorphisms. Here then is an important lesson the mathematicians learned in the twentieth century: whenever you are concerned with a structured manifold, study its group of automorphisms. Also the inverse problem, which Felix Klein stressed in his famous Erlangen program (1872), deserves attention: Given a group of transformations of a manifold σ, determine such relations or operations as are invariant with respect to the group.[5]

If in studying a group of transformations we ignore the fact that it consists of transformations and look merely at the way in which any two of its transformations S, T give rise to a composite ST, we obtain the abstract composition schema of the group. Hence an *abstract group* is a set of elements (of unknown or irrelevant nature) for which an operation of composition is defined generating an element st from any two elements s, t such that the following axioms hold:

1. There is a unit element e such that $es = se = s$ for every s.
2. Every element s has an inverse s^{-1} such that $ss^{-1} = s^{-1}s = e$.
3. The associative law $(st)u = s(tu)$ holds.

It is perhaps the most astonishing experience of modern mathematics how rich in consequences these three simple axioms are. A realization of an abstract group by transformations of a given manifold σ is obtained by associating with every element s of the group a transformation S of σ, $s \to S$, such that $s \to S, t \to T$ imply $st \to ST$. In general, the commutative law $st = ts$ will not hold. If it does, the group is called commutative or Abelian (after the Norwegian mathematician Niels Henrik Abel). Because composition of group elements in general does not satisfy the commutative law, it has proved convenient to use the term "ring" in the wider sense in which it does not imply the commutative law for multiplication. (However, in speaking of a field one usually assumes this law.)

The simplest mappings are the linear ones. They operate in a vector space. The vectors in our ordinary three-dimensional space are directed segments AB leading from a point A to a point B. The vector AB is considered equal to $A'B'$ if a parallel displacement (translation) carries AB into $A'B'$. In consequence of this convention one can add vectors and one can also multiply a vector by a number (integral, rational or even real). Addition satisfies the same axioms as enumerated for numbers in the table **T**, and it is also easy to formulate the axioms for the second operation. These axioms constitute the general axiomatic notion of vector space, which is therefore an algebraic and not a geometric concept. The numbers which serve as multipliers of the vectors may be the elements of any ring; this generality is actually required in the application of the axiomatic vector concept to topology. However, here we shall assume that they form a field. Then one sees at once that one can ascribe to the vector space a natural number n as its dimensionality in this sense: there exist n vectors $e_1, \cdots, e_n$ such that every vector may be expressed in one and only one way as a linear combination $x_1 e_1 + \cdots + x_n e_n$, where the "coordinates" x_i are definite numbers of the field. In our three-dimensional space n equals 3, but mechanics and physics give ample occasion to use the general algebraic notion of an n-dimensional vector space for higher n.

The endomorphisms of a vector space are called its *linear mappings*; as such they allow composition ST (perform first the mapping S, then T), but they also allow addition and multiplication by numbers γ: if S sends the arbitrary vector x into xS, T into xT, then $S+T, \gamma S$ are those linear mappings which send x into $(xS) + (xT)$ and $\gamma \cdot xS$, respectively. We must forego to describe how in terms of a vector basis $e_1, \cdots, e_n$ a linear mapping is represented by a square matrix of numbers.

Often rings occur — they are then called *algebras* — which are at the same time vector spaces, i.e., for which three operations, addition of two elements, multiplication of two elements and multiplication of an element by a number, are defined in such manner as to satisfy the characteristic axioms. The linear mappings of an n-dimensional vector space themselves form such an algebra, called the complete matric algebra (in n dimensions). According to quantum mechanics the observables of a physical system form an algebra of special type with a non-commutative multiplication. In the hands of the physicists abstract algebra has thus become a key that unlocked to them the secrets of the atom. A realization of an abstract group by linear transformations of a vector space is called *representation*. One may also speak of representations of a ring or an algebra: in each case the representation can be described as a homomorphic mapping of the group or ring or algebra into the complete matric algebra (which indeed is a group and a ring and an algebra, all in one).

5. Finale.

After spending so much time on the explanation of the notions I can be brief in my enumeration of some of the essential achievements for which they provided the tools. If g is a subgroup of the group G, one may identify elements s, t of G that are congruent mod g, i.e., for which st^{-1} is in g; g is a "self-conjugate" subgroup if this process of identification carries G again into a group, the "factor group" G/g. The group-theoretic core of Galois' theory is a theorem due to C. Jordan and O. Hölder which deals with the several ways in which one may break down a given finite group G in steps $G = G_0, G_1, G_2, \cdots$, each G_i being a self-conjugate subgroup of the preceding group G_{i-1}. Under the assumption that this is done in as small steps as possible, the theorem states, the steps (factor groups) $G_{i-1}/G_i(i = 1, 2, \dots)$ in one such "composition series" are isomorphic to the steps, suitably rearranged, in a second such series. The theorem is very remarkable in itself, but perhaps the more so as its proof rests on the same argument by which one proves what I consider the most fundamental proposition in all mathematics, namely the fact that if you count a finite set of elements in two ways, you end up with the same number n both times. The Jordan-Hölder theorem in recent times received a much more natural and general formulation by (1) abandoning the assumption that the breaking down is done in the smallest possible steps, and (2) by admitting only such subgroups as are invariant with respect to a given set of endomorphic mappings of G. It thus has become applicable to infinite as well as finite groups, and provided a common denominator for quite a number of important algebraic facts.

The theory of representations of finite groups, the most systematic and substantial part of group theory developed shortly before the turn of the century by G. Frobenius, taught us that there are only a few irreducible representations, of which all others are composed. This theory was greatly simplified after 1900 and later carried over, first to continuous groups that have the topological property of compactness, but then also to all infinite groups, with a restrictive imposition (called almost-periodicity) on the representations. With these generalizations one trespasses the limits of algebra, and a few more words will have to be said about it under the title analysis. New phenomena occur if representations of finite groups in fields of prime characteristic are taken into account, and from their investigation profound number-theoretic consequences have been derived. It is easy to embed a finite group into an algebra, and hence facts about representations of a group are best deduced from those of the embedding algebra. At the beginning of the century algebras seemed to be ferocious beasts of unpredictable behavior, but after fifty years of investigation they, or at least the variety called semi-simple, have become remarkably tame; indeed the wild things do not happen in these superstructures, but in the underlying commutative "number" fields. In the nineteenth century geometry seemed to have been reduced to a study of invariants of groups; Felix Klein formulated this standpoint explicitly in his Erlangen program. But the full linear group was practically the only group whose invariants were studied. We have now outgrown this limitation and no longer ignore all the other continuous groups one encounters in algebra, analysis, geometry and physics. Above all we have come to realize that the theory of invariants has to be subsumed under that of representations. Certain infinite discontinuous groups, like the unimodular and the modular groups, which are of special importance to number theory, witness Gauss' class theory of quadratic forms, have been studied with remarkable success and profound results. The macroscopic and microscopic symmetries of crystals are described by discontinuous groups of motions, and it has been proved for n dimensions, what had long been known for 3 dimensions, that in a certain sense there is but a finite number of possibilities for these crystallographic groups. In the nineteenth century Sophus Lie had reduced a continuous group to its "germ" of infinitesimal elements. These elements form a sort of algebra in which the associative law is replaced by a different type of law. A Lie algebra is a purely algebraic structure, especially if the numbers which act as multipliers are taken from an algebraically defined field rather than from the continuum of real numbers Ω. These Lie groups have provided a new playground for our algebraists.

The constructions of the mathematical mind are at the same time free and necessary. The individual mathematician feels free to define his notions and to set up his axioms as he pleases. But the question is, will he get his fellow mathematicians interested in the constructs of his imagination. We can not help feeling that certain mathematical structures which have evolved through the combined efforts of the mathematical community bear the stamp of a necessity not affected by the accidents of their historical birth. Everybody who looks at the spectacle of modern algebra will be struck by this complementarity of freedom and necessity.

Part II. Analysis. Topology. Geometry. Foundations.

6. Linear operators and their spectral decomposition. Hilbert space.

A mechanical system of n degrees of freedom in stable equilibrium is capable of oscillations deviating "infinitely little" from the state of equilibrium. It is a fact of fundamental significance not only for physics but also for music that all these oscillations are superpositions of n "harmonic" oscillations with definite frequencies. Mathematically the problem of determining the harmonic oscillations amounts to constructing the principal axes of an ellipsoid in an n-dimensional Euclidean space. Representing the vectors x in this space by their coordinates $(x_1, x_2, \cdots, x_n)$ one has to solve an equation

$$x - \lambda \cdot Kx = 0,$$

where K denotes a given linear operator (= linear mapping); λ is the square of the unknown frequency ν of the harmonic oscillation, whereas the "eigenvector" x characterizes its amplitude. Define the scalar product (x, y) of two vectors x and y by the sum $x_1 y_1 + \cdots + x_n y_n$. Our "affine" vector space is made into a metric one by assigning to any vector x the length $\|x\|$ given by $\|x\|^2 = (x, x)$, and this metric is the Euclidean one so familiar to us from the 3-dimensional case and epitomized by the "Pythagoras."[6] The linear operator K is symmetric in the sense that $(x, Ky) = (Kx, y)$. The field of numbers in which we operate here is, of course, the continuum of all real numbers. Determination of the n frequencies ν or rather of the corresponding eigenvalues $\lambda = \nu^2$ requires the solution of an algebraic equation of degree n (often known as the secular equation, because it first appeared in the theory of the secular perturbations of the planetary system).

More important in physics than the oscillations of a mechanical system of a finite number of degrees of freedom are the oscillations of continuous media, as the mechanical-acoustical oscillations of a string, a membrane or a 3-dimensional elastic body, and the electromagnetic-optical oscillations of the "ether." Here the vectors with which one has to operate are continuous functions $x(s)$ of a point s with one or several coordinates that vary over a given domain, and consequently K is a linear *integral* operator. Take for instance a straight string of length 1, the points of which are distinguished by a parameter s varying from 0 to 1. Here (x, x) is the integral $\int_0^1 x^2(s) \cdot ds$, and the problem of harmonic oscillations (which first suggested to the early Greeks the idea of a universe ruled by harmonious mathematical laws) takes the form of the integral equation

[1]
$$x(s) - \lambda \int_0^1 K(s,t)x(t)\,dt = 0, \quad (0 \leqq s \leqq a),$$

where

[1']
$$K(s,t) = \left(\frac{a}{\pi}\right)^2 \cdot \begin{cases} s(1-t) & \text{for } s \leqq t \\ (1-s)t & \text{for } s \geqq t \end{cases}$$

and a is a constant determined by the physical conditions of the string. The solutions are

$$\lambda = (na)^2, \quad x(s) = \sin n\pi s,$$

where n is capable of all positive integral values $1, 2, 3, \cdots$. This fact that the frequencies of a string are integral multiples na of a ground frequency a is the basic law of musical harmony. If one prefers an optical to an acoustic language one speaks of the *spectrum* of eigenvalues λ.

After Fredholm at the very close of the nineteenth century had developed the theory of linear integral equations it was Hilbert who in the next decade established the general *spectral theory of symmetric linear operators* K. Only twenty years earlier it had required the greatest mathematical efforts to prove the existence of the ground frequency for a membrane, and now constructive proofs for the existence of the whole series of harmonic oscillations and their characteristic frequencies were given under very general assumptions concerning the oscillating medium. This was an event of great consequence both in mathematics and theoretical physics. Soon afterwards Hilbert's approach made it possible to establish those asymptotic laws for the distribution of eigenvalues the physicists had postulated in their statistical treatment of the thermodynamics of radiation and elastic bodies.

Hilbert observed that an arbitrary continuous function $x(s)$ defined in the interval $0 \leq s \leq 1$ may be replaced by the sequence

$$x_n = \sqrt{2} \int_0^1 x(s) \cdot \sin n\pi s \cdot ds, \quad n = 1, 2, 3, \cdots,$$

of its Fourier coefficients. Thus there is no inner difference between a vector space whose elements are functions $x(s)$ of a continuous variable and one whose elements are infinite sequences of numbers $(x_1, x_2, x_3, \cdots)$. The square of the "length," $\int_0^1 x^2(s) \cdot da$ equals $x_1^2 + x_2^2 + x_3^2 + \cdots$. Between the two forms in which one may pass from a finite sum to a limit, the infinite sum $a_1 + a_2 + a_3 + \cdots$ and the integral $\int_0^1 a(s) \cdot ds$, there is therefore here no essential difference. Thus an axiomatic formulation is called for. To the axioms for an (affine) vector space one adds the postulate of the existence of a scalar product (x, y) of any two vectors (x, y) with the properties characteristic for Euclidean metric: (x, y) is a number depending linearly on either of the two argument vectors x and y; it is symmetric, $(x, y) = (y, x)$; and $(x, x) = \|x\|^2$ is positive except for $x = 0$. The axiom of finite dimensionality is replaced by a denumerability axiom of more general character. All operations in such a space are greatly facilitated if it is assumed to be complete in the same sense that the system of real numbers is complete; i.e., if the following is true: Given a "convergent" sequence $x', x'', \cdots$ of vectors, namely, one for which $\|x^{(m)} - x^{(n)}\|$ tends to zero with m and n tending to infinity, there exists a vector a toward which this sequence converges, $\|x^{(n)} - a\| \to 0$ for $n \to \infty$. A non-complete vector space can be made complete by the same construction by which the system of rational numbers is completed to form that of real numbers. Later authors have coined the name "Hilbert space" for a vector space satisfying these axioms.

Hilbert himself first tackled only integral operators in the strict sense as exemplified by [1]. But soon he extended his spectral theory to a far wider class, that of bounded (symmetric) linear operators in Hilbert space. Boundedness of the linear operator requires the existence of a constant M such that $\|Kx\|^2 \leq M \cdot \|x\|^2$ for all vectors x of finite length $\|x\|$. Indeed the restriction to integral operators would be unnatural since the simplest operator, the identity $x \to x$, is not of this type. And now one of those events happened, unforeseeable by the wildest imagination, the like of which could tempt one to believe in a pre-established harmony between physical nature and mathematical mind: Twenty years after Hilbert's investigations *quantum mechanics* found that the observables of a physical system are represented by the linear symmetric operators in a Hilbert space and that the eigenvalues and eigenvectors of that operator which represents *energy* are the energy levels and corresponding stationary quantum states of the system. Of course this quantum-physical interpretation added greatly to the interest in the theory and led to a more scrupulous investigation of it, resulting in various simplifications and extensions.

Oscillations of continua, the boundary value problems of classical physics and the problem of energy levels in quantum physics, are not the only titles for applications of the theory of integral equations and their spectra. One other somewhat isolated application is the solution of *Riemann's monodromy problem* concerning analytic functions of a complex variable z. It concerns the determination of n analytic functions of z which remain regular under analytic continuation along arbitrary paths in the z-plane provided these avoid a finite number of singular points, whereas the functions undergo a given linear transformation with constant coefficients when the path circles one of these points.

Another surprising application is to the establishment of the fundamental facts, in particular of the completeness relation, in the theory of *representations of continuous compact groups*. The simplest such group consists of the rotations of a circle, and in that case the theory of representations is nothing but the theory of the so-called Fourier series, which expresses an arbitrary periodic function $f(s)$ of period 2π in terms of the harmonic oscillations

$$\cos ns, \quad \sin ns, \quad n = 0, 1, 2, \cdots.$$

In Nature functions often occur with hidden non-commensurable periodicities. The mathematician Harald Bohr, the brother of the physicist Niels Bohr, prompted by certain of his investigations concerning the Riemann zeta function, developed the general mathematical theory of such *almost periodic functions*. One may describe his theory as that of almost periodic representations of the simplest continuous group one can imagine, namely, the group of all translations of a straight line. His main results could be carried over to arbitrary groups. No restriction is imposed on the group, but the representations one studies are supposed to be almost-periodic. For a function $x(s)$, the argument s of which runs over the group elements, while its values are real or complex numbers, almost-periodicity amounts to the requirement that the group be compact in a certain topology induced by the function. This relative compactness instead of absolute compactness is sufficient. Even so the restriction is severe. Indeed the most important representations of the classical continuous groups are not almost-periodic. Hence the theory is in need of further extension, which has busied a number of American and Russian mathematicians during the last decade.

7. Lebesgue's integral. Measure theory. Ergodic hypothesis.

Before turning to other applications of operators in Hilbert space I must mention the, in all probability final, form given to the idea of integration by Lebesgue at the beginning of our century. Instead of speaking of the area of a piece of the 2-dimensional plane referred to coordinates x, y, or the volume of a piece of the 3-dimensional Euclidean space, we use the neutral term *measure* for all dimensions. The notions of measure and *integral* are interconnected. Any piece of space, any set of space points can be described by its characteristic function $\chi(P)$, which equals 1 or 0 according to whether the point P belongs or does not belong to the set. The measure of the point set is the integral of this characteristic function. Before Lebesgue one first defined the integral for continuous functions; the notion of measure was secondary; it required transition from continuous to such discontinuous functions as $\chi(P)$. Lebesgue goes the opposite and perhaps more natural way: for him measure comes first and the integral second. The one-dimensional space is sufficient for an illustration. Consider a real-valued function $y = f(x)$ of a real variable x which maps the interval $0 \leqq x \leqq 1$ into a finite interval $a \leqq y \leqq b$. Instead of subdividing the interval of the argument x Lebesgue subdivides the interval (a, b) of the dependent variable y into a finite number of small subintervals $a_i \leqq y < a_{i+1}$, say of lengths $< \epsilon$, and then determines the measure m_i of the set S_i on the x-axis, the points of which satisfy the inequality $a_i \leqq f(x) < a_{i+1}$. The integral lies between the two sums $\sum_i a_i m_i$ and $\sum_i a_{i+1} m_i$ which differ by less than ϵ, and thus can be computed with any degree of accuracy. In determining the measure of a point set — and this is the more essential modification — Lebesgue covers the set with infinite sequences, rather than finite ensembles, of intervals. Thus, to the set of rational x in the interval $0 \leqq x \leqq 1$ no measure could be ascribed before Lebesgue. But these rational numbers can be arranged in a denumerable sequence $a_1, a_2, a_3, \cdots$, and, after choosing a positive number ϵ as small as one likes, one can surround the point a_n by an interval of length $\epsilon/2^n$ with the center a_n. Thus the whole set of rational points is enclosed in a sequence of intervals of total length

$$\epsilon(1/2 + 1/2^2 + 1/2^3 + \cdots) = \epsilon;$$

and according to Lebesgue's definition its measure is therefore less than (the arbitrary positive) ϵ and hence zero. The notion of *probability* is tied to that of measure, and for this reason mathematical statisticians are deeply interested in measure theory. Lebesgue's idea has been generalized in several directions. The two fundamental operations one can perform with sets are: forming the intersection and the union of given sets, and thus sets may be considered as elements of a "*Boolean algebra*" with these two operations, the properties of which may be laid down in a number of axioms resembling the arithmetical axioms for addition and multiplication. Hence one of the questions which has occupied the more axiomatically minded among the mathematicians and statisticians is concerned with the introduction of measure in abstract Boolean algebras.

Lebesgue's integral is important in our present context, because those real-valued functions $f(x)$ of a real variable x ranging over the interval $0 \leqq x \leqq 1$, the squares of which are Lebesgue-integrable, form a complete Hilbert space — provided two functions $f(x), g(x)$ are considered equal if those values x for which $f(x) \neq g(x)$ form a set of measure zero (Riesz-Fischer thesorem).

The mechanical equations for a system of n degrees of freedom in Hamilton's form uniquely determine the state tP at the moment t if the state P at the moment $t = 0$ is given. Such is the precise formulation of the law of causality in mechanics. The possible states P form the points of a $(2n)$-dimensional phase space, and for a fixed t and an arbitrary P the transition $P \to tP$ is a measure-preserving mapping (t). These transformations form a group: $(t_l)(t_2) = (t_1 + t_2)$. For a given P and a variable t the point tP describes the consecutive states which this system assumes if at the moment $t = 0$ it is in the state P. Considering P as a particle of a $(2n)$-dimensional fluid which fills the phase-space and ascribing to the particle P the position tP at the time t, one obtains the picture of an incompressible fluid in stationary flow. The statistical derivation of the laws of thermodynamics makes use of the so-called *ergodic hypothesis* according to which the path of an arbitrary individual particle P (excepting initial states P which form a set of measure zero) covers the phase-space (or at least that $(2n - 1)$-dimensional sub-space of it where the energy has a given value) everywhere dense, so that in the course of its history the probability of finding it in this or that part of the space is the same for any parts of equal measure. Nineteenth century mathematics seemed to be a long way off from proving this hypothesis with any degree of generality. Strangely enough it was proved shortly after the transition from classical to quantum mechanics had rendered the hypothesis almost valueless, and it was proved by making use of the mathematical apparatus of quantum physics. Under the influence of the mapping (t), $P \to tP$, any function $f(P)$ in phase-space is transformed into the function $f' = U_i f$, defined by the equation $f'(tP) = f(P)$. The U_i form a group of operators in the Hilbert space of arbitrary functions $f(P)$, $U_{t_1} U_{t_2} = U_{t_1 + t_2}$, and application of spectral decomposition to this group enabled J. von Neumann to deduce the ergodic hypothesis with two provisos: (1) Convergence of a sequence of functions $f_n(P)$ toward a function $f(P)$, $f_n \to f$, is understood (as it would in quantum mechanics, namely) as convergence in Hilbert space where it means that the total integral of $(f_n - f)^2$ tends to zero with $n \to \infty$; (2) one assumes that there are no subspaces of the phase-space which are invariant under the group of transformations (t) except those spaces that are in Lebesgue's sense equal either to the empty or the total space. Shortly afterwards proofs were also given for other interpretations of the notion of convergence.

The laws of nature can either be formulated as differential equations or as "principles of variation" according to which certain quantities assume extremal values under given conditions. For instance, in an optically homogeneous or non-homogeneous medium the light travels along that road from a given point A to a given point B for which the time of travel assumes minimal value. In potential theory the quantity which assumes a minimum is the so-called Dirichlet integral. Attempts to establish directly the existence of a minimum had been discouraged by Weierstrass' criticism in the nineteenth century. Our century, however, restored the direct methods of the Calculus of Variation to a position of honor after Hilbert in 1900 gave a direct proof of the Dirichlet principle and later showed how it can be applied not only in establishing the fundamental facts about functions and integrals ("algebraic" functions and "abelian" integrals) on a compact Riemann surface (as Riemann had suggested 50 years earlier) but also for deriving the basic propositions of the *theory of uniformization*. That theory occupies a central position in the theory of functions of one complex variable, and the first decade of the 20th century witnessed the first proofs by P. Koebe and H. Poincaré of these propositions conjectured about 25 years before by Poincaré himself and by Felix Klein. As in an Euclidean vector space of finite dimensionality, so in the Hilbert space of infinitely many dimensions, this fact is true: Given a linear (complete) subspace E, any vector may be split in a uniquely determined manner into a component lying in E (orthogonal projection) and one perpendicular to E. Dirichlet's principle is nothing but a special case of this fact. But since the function-theoretic applications of orthogonal projection in Hilbert space which we alluded to are closely tied up with topology we had better turn first to a discussion of this important branch of modern mathematics: topology.

8. Topology and harmonic integrals

Essential features of the modern approach to *topology* can be brought to light in its connection with the, only recently developed, theory of *harmonic integrals*. Consider a stationary magnetic field h in a domain G which is free from electric currents. At every point of G it satisfies two differential conditions which in the usual notations of vector analysis are written in the form div $h = 0$, rot $h = 0$.[7] A field of this type is called harmonic. The second of these conditions states that the line integral of h along a closed curve (cycle) $C, \int_C h$, vanishes provided C lies in a sufficiently small neighborhood of an arbitrary point of G. This implies $\int_C h = 0$ for any cycle C in G that is the boundary of a surface in G. However, for an arbitrary cycle C in G the integral is equal to the electric current surrounded by C.

Let the phrase "C homologous to zero," $C \sim 0$, indicate that the cycle C in G bounds a surface in G. One can travel over a cycle C in the opposite sense, thus obtaining $-C$, or travel over it $2, 3, \cdots$ times, thus obtaining $2C, 3C, \cdots$; and cycles may be added and subtracted from each other (if one does not insist that cycles are of one piece). Two cycles C, C' are called homologous, $C \sim C'$, if $C - C' \sim 0$. Note that $C \sim 0, C' \sim 0$ imply $-C \sim 0, C + C' \sim 0$. Hence the cycles form a commutative group under addition, the "Betti group," if homologous cycles are considered as one and the same group element. These notions of *cycles and their homologies* may be carried over from a three-dimensional domain in Euclidean space to any n-dimensional manifold, in particular to closed (compact) manifolds like the two-dimensional surfaces of the sphere or the torus; and on an n-dimensional manifold we can speak not only of 1-dimensional, but also of 2-, 3-, $\cdots$, n-dimensional cycles. The notion of a harmonic vector field permits a similar generalization, harmonic tensor field (harmonic form) of rank r $(r = 1, 2, \cdots, n)$, provided the manifold bears a Riemannian metric, an assumption the meaning of which will be discussed later in the section on geometry. Any tensor field (linear differential form) of rank r may be integrated over an r-dimensional cycle.

The fundamental problem of *homology theory* consists in determining the structure of the Betti group, not only for 1-, but also for 2-, $\cdots$, n-dimensional cycles, in particular in determining the number of linearly independent cycles (Betti number). (ν cycles $C_1, \cdots, C_\nu$, are linearly independent if there exists no homology $k_1 C_1 + \cdots + k_\nu C_\nu \sim 0$ with integral coefficients k except $k_1 = \cdots = k_\nu = 0$.) The fundamental theorem for harmonic forms on compact manifolds states that, given ν linearly independent cycles $C_1, \cdots, C_\nu$, there exists a harmonic form h with pre-assigned periods

$$\int_{C_1} h = \pi_1, \cdots, \qquad \int_{C_\nu} h = \pi_\nu.$$

H. Poincaré developed the algebraic apparatus necessary to formulate exactly the notions of cycle and homology. In the course of the twentieth century it turned out that in most problems co-homologies are easier to handle than homologies. I illustrate this for 1-dimensional cycles. A line C_1 leading from a point p_1 to p_2, when followed by a line C_2 leading from p_2 to a third point p_3, gives rise to a line $C_1 + C_2$ leading from p_l to p_3. The line integral $\int_C h$ of a given vector field h along an arbitrary (closed or open) line C is an additive function $\phi(C)$ of C, $\phi(C_1 + C_2) = \phi(C_1) + \phi(C_2)$. If moreover rot h vanishes everywhere, then $\phi(C) = 0$ for any closed line C that lies in a sufficiently small neighborhood of a point, whatever this point may be. Any real-valued function ϕ satisfying these two conditions may be called an abstract integral. The cohomology $\phi \sim 0$ means that $\phi(C) = 0$ for any closed line C, and thus it is clear what the cohomology $k_1 \phi_1 + \cdots + k_\nu \phi_\nu \sim 0$ with arbitrary real coefficients $k_1, \cdots, k_\nu$ means. The homology $C \sim 0$ could now be defined, not by the condition that the cycle C bounds, but by the requirement that $\phi(C) = 0$ for every abstract integral ϕ. With the convention that any two abstract integrals ϕ, ϕ' are identified if $\phi - \phi' \sim 0$, these integrals form a vector space, and the dimensionality of this vector space is now introduced as the Betti number. And the fundamental theorem for harmonic integrals on a compact manifold now asserts that for any given abstract integral ϕ there exists one and only one harmonic vector field h whose integral is cohomologous to ϕ, $\int_C h = \phi(C)$, for every cycle C (realization of the abstract integral in concreto by a harmonic integral).

J. W. Alexander discovered an important result connecting the Betti numbers of a manifold M that is embedded in the n-dimensional Euclidean space R_n with the Betti numbers of the complement $R_n - M$ (*Alexander's duality theorem*).

The difficulties of topology spring from the double aspect under which one can consider continuous manifolds. Euclid looked upon a figure as an assemblage of a finite number of geometric elements, like points, straight lines, circles, planes, spheres. But after replacing each line or surface by the set of points lying on it one may also adopt the set-theoretic view that there is only one sort of elements, points, and that any (in general infinite) set of points can serve as a figure. This modern standpoint obviously gives geometry far greater generality and freedom.

In topology, however, it is not necessary to descend to the points as the ultimate atoms, but one can construct the manifold like a building from "blocks" or cells, and a finite number of such cells serving as units will do, provided the manifold is compact. Thus it is possible here to revert to a treatment in Euclid's "finitistic" style (combinatorial topology).

On the first standpoint, manifold as a point set, the task is to formulate that *continuity* by which a point p approaching a given point p_0 becomes gradually indistinguishable from p_0. This is done by associating with p_0 the *neighborhoods* of p_0, an infinite shrinking sequence of sub-sets $U_1 \supset U_2 \supset U_3 \supset \cdots$, all containing p_0. ($U \supset V$ means: the set U contains V.) For example, in a plane referred to Cartesian coordinates x, y we may choose as the nth neighborhood U of a point $p_0 = (x_0, y_0)$ the interior of the circle of radius $1/2^n$ around p_0. The notion of convergence, basic for all continuity considerations, is defined in terms of the sequence of neighborhoods as follows: A sequence of points $p_1, p_2, \cdots$ converges to p_0 if for every natural number n there is an N so that all points p_ν with $\nu > N$ lie in the nth neighborhood U_n of p_0. Of course, the choice of the neighborhoods U_n is arbitrary to a certain extent. For instance, one could also have chosen as the nth neighborhood V_n of (x_0, y_0) the square of side $2/n$ around (x_0, y_0), to which a point (x, y) belongs, if

$$-1/n < x - x_0 < 1/n, \qquad -1/n < y - y_0 < 1/n.$$

However the sequence V_n is equivalent to the sequence U_n in the sense that for every n there is an n' such that $U_{n'} \subset V_n$ (and thus $U_\nu \subset V_n$ for $\nu \geqq n'$), and also for every m an m' such that $V_{m'} \subset U_m$; and consequently the notion of convergence for points is the same, whether based on the one or the other sequence of neighborhoods. It is clear how to define continuity of a mapping of one manifold into another. A one-to-one mapping of two manifolds upon each other is called topological if continuous in both directions, and two manifolds that can be mapped topologically upon each other are topologically equivalent. Topology investigates such properties of manifolds as are invariant with respect to topological mappings (in particular with respect to continuous deformations).

A continuous function $y = f(x)$ may be approximated by piecewise linear functions. The corresponding device in higher dimensions, the method of *simplicial approximations* of a given continuous mapping of one manifold into another, is of great importance in set-theoretic topology. It has served to develop a general *theory of dimensions*, to prove the topological invariance of the Betti groups, to define the decisive notion of the degree of mapping (*"Abbildungsgrad,"* L. E. J. Brouwer) and to prove a number of interesting fixed point theorems. For instance, a continuous mapping of a square into itself has necessarily a fixed point, i.e., a point carried by the mapping into itself. Given two continuous mappings of a (compact) manifold M into another M', one can ask more generally for which points p on M both images on M' coincide. A famous formula by S. Lefschetz relates the "total index" of such points with the homology theory of cycles on M

and M'.

Application of fixed point theorems to functional spaces of infinitely many dimensions has proved a powerful method to establish the existence of solutions for non-linear differential equations. This is particularly valuable, because the hydrodynamical and aerodynamical problems are almost all of this type.

Poincaré found that a satisfactory formulation of the homology theory of cycles was possible only from the second standpoint where the n-dimensional manifold is considered as a conglomerate of n-dimensional cells. The boundary of an n-dimensional cell (n-cell) consists of a finite number of $(n-1)$-cells, the boundary of an $(n-1)$-cell consists of a finite number of $(n-2)$-cells, etc. The *combinatorial skeleton* of the manifold is obtained by assigning symbols to these cells and then stating in terms of their symbols which $(i-1)$-cells belong to the boundary of any of the occurring i-cells ($i = 1, 2, \cdots, n$). From the cells one descends to the points of the manifold by a repeated process of sub-division which catches the points in an ever finer net. Since this subdivision proceeds according to a fixed combinatorial scheme, the manifold is in topological regard completely fixed by its combinatorial skeleton. And at once the question arises under what circumstances two given combinatorial skeletons represent the same manifold, i.e., lead by iterated sub-division to topologically equivalent manifolds. We are far from being able to solve this fundamental problem. Algebraic topology, which operates with the combinatorial skeletons, is in itself a rich and beautiful theory, linked in various ways with the basic notions and theorems of algebra and group theory.

The connection between algebraic and set-theoretic topology is fraught with serious difficulties which are not yet overcome in a quite satisfactory manner. So much, however, seems clear that one had better start, not with a division into cells, but with a covering by patches which are allowed to overlap. From such a pattern the fundamental topologically invariant concepts are to be developed. The above notion of an abstract integral, which relates homology and cohomology, is indication; it can indeed be used for a direct proof of the invariance of the first Betti number without the tool of simplicial approximation.

9. Conformal mapping, meromorphic functions, Calculus of Variation in the large.

Homology theory, in combination with the Dirichlet principle or the method of orthogonal projection in Hilbert space, leads to the theory of harmonic integrals, in particular for the lowest dimension $n = 2$ to the theory of abelian integrals on Riemann surfaces. But for Riemann surfaces the Dirichlet principle also yields the fundamental facts concerning uniformization of analytic functions of one variable if one combines it with the *homotopy* (not homology) *theory* of closed curves. Whereas a cycle is homologous to zero if it bounds, it is homotopic to zero if it can be contracted into a point by continuous deformation. The homotopy theory of 1- and more-dimensional cycles has recently come to the fore as an important branch of topology, and the group-theoretic aspect of homotopy has led to some surprising discoveries in abstract group theory. Homotopy of 1-dimensional cycles is closely related with the idea of the *universal covering manifold* of a given manifold. Given a continuous mapping $p \to p'$ of one manifold M into another M', the point p' may be considered as the trace or projection in M' of the arbitrary point p on M, and thus M becomes a manifold covering M'. There may be no point or several points p on M which lie over a given point p' of M' (which are mapped into p'). The mapping is without ramifications if for any point p_0 of M it is one-to-one (and continuous both ways) in a sufficiently small neighborhood of p_0. Let p_0 be a point on M, p'_0 its trace on M', and C' a curve on M' beginning at p'_0. If M covers M' without ramifications we can follow this curve on M by starting at p_0, at least up to a certain point where we would run against a "boundary of M relative to M'." Of chief interest are those covering manifolds M over a given M' for which this never happens and which therefore cover M' without ramifications and relative boundaries. The best way of defining the central topological notion "simply connected" is by describing a simply connected manifold as one having no other unramified unbounded covering but itself. There is a strongest of all unramified unbounded covering manifolds, the universal covering manifold, which can be described by the statement that on it a curve C is closed only if its trace C' is (closed and) homotopic to zero. The proof of the fundamental theorem on uniformization consists of two parts: (1) constructing the universal covering manifold of the given Riemann surface, (2) constructing by means of the Dirichlet principle a one-to-one conformal mapping of the covering manifold upon the interior of a circle of finite or infinite radius.

All we have discussed so far in our account of analysis, is in some way tied up with operators and projections in Hilbert space, the analogue in infinitely many dimensions of Euclidean space. In H. Minkowski's *Geometry of Numbers* [1968] distances $|AB|$, which are different from the Euclidean distance but satisfy the axioms that $|BA| = |AB|$ and that in a triangle ABC the inequality $|AC| \leq |AB| + |BC|$ holds, were used to great advantage for obtaining numerous results concerning the solvability of inequalities by integers. We do not find time here to report on the progress of this attractive branch of number theory during the last fifty years. In infinitely many dimensions spaces endowed with a metric of this sort, of a more general nature than the Euclid-Hilbert metric, have been introduced by Banach, not however for number-theoretic but for purely analytic purposes. Whether the importance of the subject justifies the large number of papers written on *Banach spaces* is perhaps questionable.

The Dirichlet principle is but the simplest example of the direct methods of the Calculus of Variation as they came into use with the turn of the century. It was by these methods that the theory of *minimal surfaces*, so closely related to that of analytic functions, was put on a new footing. What we know about non-linear differential equations has been obtained either by the topological fixed point method (see above) or by the so-called continuity method or by constructing their solutions as extremals of a suitable functional.

A continuous function on an n-dimensional compact manifold assumes somewhere a minimum and somewhere else a maximum value. Interpret the function as altitude. Besides summit (local maximum) and bottom (local minimum) one has the further possibility of a saddle point (pass) as a point of "stationary" altitude. In n dimensions the several possibilities are indicated by an inertial index k which is capable of the values $k = 0, 1, 2, \cdots, n$, the value $k = 0$ a minimum and $k = n$ characterizing a maximum. Marston Morse discovered the inequality $M_k \geqq B_k$ between the number M_k of stationary points of index k and the Betti number B_k of linearly independent homology classes of k-dimensional cycles. In their generalization to functional spaces these relations have opened a line of study adequately described as *Calculus of Variation in the large*.

Development of the theory of uniformization for analytic functions led to a closer investigation of *conformal mapping* of 2-dimensional manifolds in the large, which resulted in a number of theorems of surprising simplicity and beauty. In the same field there is to register an enormous extension of our knowledge of the behavior of *meromorphic functions*, i.e., single-valued analytic functions of the complex variable z which are regular everywhere with the exception of isolated "poles" (points of infinity). Towards the end of the previous century Riemann's zeta function had provided the stimulus for a deeper study of "entire functions" (functions without poles). The greatest stride forward, both in methods and results, was marked by a paper on meromorphic functions published in 1925 by the Finnish mathematician Rolf Nevanlinna. Besides meromorphic functions in the z-plane one can study such functions on a given Riemann surface; and in the way in which the theory of algebraic functions (equal to meromorphic functions on a compact Riemann surface) as a theory of algebraic curves in two complex dimensions may be generalized to any number of dimensions, so one can pass from meromorphic functions to meromorphic curves.

The theory of *analytic functions of several complex variables*, in spite of a number of deep results, is still in its infancy.

10. Geometry.

After having dealt at some length with the problems of analysis and topology I must be brief about geometry. Of subjects mentioned before, minimal surfaces, conformal mapping, algebraic manifolds and the whole of topology could be subsumed under the title of geometry. In the domain of *elementary axiomatic geometry* one strange discovery, that of von Neumann's pointless "continuous geometries" stands out, because it is intimately interrelated with quantum mechanics, logic, and the general algebraic theory of "lattices." The 1-, 2-, $\cdots, n$-dimensional linear manifolds of an n-dimensional vector space form the 0-, 1-, $\cdots, (n-1)$-dimensional linear manifolds in an $(n-1)$-dimensional projective point space. The usual axiomatic foundation of projective geometry uses the points as the primitive elements or atoms of which the higher-than-zero-dimensional manifolds are composed. However, there is possible a treatment where the linear manifolds of all dimensions figure as elements, and the axioms deal with the relation "B contains A" ($A \subset B$) between these elements and the operation of intersection, $A \cap B$, and of union, $A \cup B$, performed on them; the union $A \cup B$ consists of all sums $x + y$ of a vector x in A and a vector y in B. In quantum logic this relation and these operations correspond to the relation of implication ("The statement A implies B") and the operations 'and,' 'or' in classical logic. But whereas in classical logic the distributive law

$$A \cap (B \cup C) = (A \cap B) \cup (A \cap C)$$

holds, this is not so in quantum logic; it must be replaced by the weaker axiom:

$$\text{If } C \subset A \text{ then } A \cap (B \cup C) = (A \cap B) \cup C.$$

On formulating the axioms without the implication of finite dimensionality one will come across several possibilities; one leads to the Hilbert space in which quantum mechanics operates, another to von Neumann's continuous geometry with its continuous scale of dimensions, in which elements of arbitrarily low dimensions exist but none of dimension zero.

The most important development of geometry in the twentieth century took place in differential geometry and was stimulated by general relativity, which showed that the world is a 4-dimensional manifold endowed with a Riemannian metric. A piece of an n-dimensional manifold can be mapped in one-to-one continuous fashion upon a piece of the n-dimensional "arithmetical space" which consists of all n-tuples $(x_l, x_2, \cdots, x_n)$ of real numbers x_i. A *Riemann metric* assigns to a line element which leads from the point $P = (x_1, \cdots, x_n)$ to the infinitely near point $P' = (x_1 + dx_1, \cdots, x_n + dx_n)$ a distance ds the square of which is a quadratic form of the relative coordinates dx_i,

$$ds^2 = \sum g_{ij} dx_i dx_j, \qquad (i, j = 1, \cdots, n)$$

with coefficients g_{ij} depending on the point P but not on the line element. This means that, in the infinitely small, Pythagoras' theorem and hence Euclidian geometry are valid, but in general not in a region of finite extension. The line elements at a point may be considered as the infinitesimal vectors of an n-dimensional vector space in P, the tangent space or the compass at P; indeed an arbitrary (differentiable) transformation of the coordinates x_i induces a linear transformation of the components dx_i of any line element at a given point P. As Levi-Civita found in 1915 the development of Riemannian geometry hinges on the fact that a Riemannian metric uniquely determines an infinitesimal parallel displacement of the vector compass at P to any infinitely near point P'. From this a general scheme for differential geometry arose in which each point

P of the manifold is associated with a homogeneous space Σ_P described by a definite group of "automorphisms," this space now taking over the role of the tangent space (whose group of automorphisms consists of all non-singular linear transformations). One assumes that one knows how this associated space Σ_P is transferred by infinitesimal displacement to the space $\Sigma_{P'}$ associated with any infinitely near point P'. The most fundamental notion of Riemannian geometry, that of curvature, which figures so prominently in Einstein's equations of the gravitational field, can be carried over to this general scheme. Thus one has erected general differential affine, projective, conformal, geometries, etc. One has also tried by their structures to account for the other physical fields existing in nature beside the gravitational one, namely the electromagnetic field, the electronic wave-field and further fields corresponding to the several kinds of elementary particles.[8] But it seems to the author that so far all such speculative attempts of building up a unified field theory have failed. There are very good reasons for interpreting gravitation in terms of the basic concepts of differential geometry. But it is probably unsound to try to "geometrize" all physical entities.

Differential geometry in the large is an interesting field of investigation which relates the differential properties of a manifold with its topological structure. The schema of differential geometry explained above with its associated spaces Σ_P and their displacements has a purely topological kernel which has recently developed under the name of *fibre spaces* into an important topological technique.

Our account of progress made during the last fifty years in analysis, geometry and topology had to touch on many special subjects. It would have failed completely had it not imparted to the reader some feeling of the close relationship connecting all these mathematical endeavors. As the last example of fibre spaces (beside many others) shows, this unity in diversity even makes a clear-cut division into analysis, geometry, topology (and algebra) practically impossible.

11. Foundations.

Finally a few words about the *foundations of mathematics*. The nineteenth century had witnessed the critical analysis of all mathematical notions including that of natural numbers to the point where they got reduced to pure logic and the ideas "set" and "mapping." At the end of the century it became clear that the unrestricted formation of sets, sub-sets of sets, sets of sets etc., together with an unimpeded application to them as to the original elements of the logical quantifiers "there exists" and "all" (cf. the sentences: the (natural) number n is even if there exists a number x such that $n = 2x$; it is odd if n is different from $2x$ for all x) inexorably leads to antinomies. The three most characteristic contributions of the twentieth century to the solution of this Gordian knot are connected with the names of L. E. J. Brouwer, David Hilbert and Kurt Gödel. Brouwer's critique of "mathematical existentialism" not only dissolved the antinomies completely but also destroyed a good part of classical mathematics that had heretofore been universally accepted.

If only the historical event that somebody has succeeded in constructing a (natural) number n with the given property P can give a right to the assertion that "there exists a number with that property" then the alternative that there either exists such a number or that all numbers have the opposite property non-P is without foundation. The principle of excluded middle for such sentences may be valid for God who surveys the infinite sequence of all natural numbers, as it were, with one glance, but not for human logic. Since the quantifiers "there is" and "all" are piled upon each other in the most manifold way in the formation of mathematical propositions, Brouwer's critique makes almost all of them meaningless, and therefore Brouwer set out to build up a new mathematics which makes no use of that logical principle. I think that everybody has to accept Brouwer's critique who wants to hold on to the belief that mathematical propositions tell the sheer truth, truth based on evidence. At least Brouwers' opponent, Hilbert, accepted it tacitly. He tried to save classical mathematics by converting it from a system of meaningful propositions into a game of meaningless formulas, and by showing that this game never leads to two formulas, F and non-F, which are inconsistent. Consistency, not truth, is his aim. His attempts at proving consistency revealed the astonishingly complex logical structure of mathematics. The first steps were promising indeed. But then Gödel's discovery cast a deep shadow over Hilbert's enterprise. Consistency itself may be expressed by a formula. What Gödel showed was this: If the game of mathematics is actually consistent then the formula of consistency cannot be proved within this game. How can we then hope to prove it at all?

This is where we stand now. It is pretty clear that our theory of the physical world is not a description of the phenomena as we perceive them, but is a bold symbolic construction. However, one may be surprised to learn that even mathematics shares this character. The success of the anti-phenomenological constructive method is undeniable. And yet the ultimate foundations on which it rests remain a mystery, even in mathematics.

Notes

[Published as Weyl 1951, *WGA* 4:464–494.]

[1] [Just this approach has been taken by Gray 2000.]
[2] [These numbers are known today as p-adic, a term Weyl himself goes on to use. See Gouvêa 1997 for a general introduction.]
[3] [See Diamond 1982 and Goldfeld 2003.]
[4] [See Vaughan and Wooley 2002.]
[5] [For the Erlangen program, see 66, above.]

[6] [By "the 'Pythagoras'," Weyl means the Pythagorean theorem, the expression of the distance as the sum of the squares of the coordinates.]

[7] [The notation curl h is currently more common than rot h.]

[8] [Weyl here alludes to his own work to geometrize electromagnetism within the framework of general relativity; see his 1952a, 282–312, 2009a, 20–24.]

Axiomatic Versus Constructive Procedures in Mathematics

(1953)

In his charming little book *A Mathematician's Apology* the late G. H. Hardy said: "It is a melancholy experience for a professional mathematician to find himself writing *about* mathematics. The function of a mathematician is to *do* something and not to talk about what he and other mathematicians have done." And a little later he continues: "I write about mathematics, because like any other mathematician who has passed sixty I have no longer the freshness of mind, the energy or patience to carry on effectively with my proper job." The mood which Hardy's words reflect with such obvious sincerity is not alien to me who long ago passed sixty, and I agree wholeheartedly with him that "mathematics is a young man's game." What I do not share is his contempt for the man who "talks about." It seems to me that in the intellectual life of men two spheres can be distinguished, the one that of doing, shaping, constructing, creating something, in which the active artist, scientist, technician, statesman move, the other that of reflection where the meaning of all this activity is questioned and which one may consider the proper domain of the philosopher. Creative activity, if not supervised by reflection, is in danger of running away from all meaning, of losing contact and perspective, of degenerating into routine, whereas the peril of reflection is that it becomes mere irresponsible "talk about," thus paralyzing the creative power of man. Philosophical reflection ought to be combined with historical reflection, the ever new appropriation and transformation of the past in the light of today's tasks. Certainly the scientist's aim and chief interest is objective truth, truth that is raised above, and independent of, the frailties and barriers of our human existence. But he cannot deny that on the other hand actual science is a branch of human endeavor and as such essentially historic like the mind's life on this earth from which it springs. To the results which it produces we must not betray sovereign life itself. Sterility would be the punishment. Day by day we all live in this tension of what is historical-human on the one, eternal-objective on the other side.

Mathematics has been called the science of the *infinite*. Indeed, the mathematician invents finite constructions by which questions are decided that by their very nature refer to the infinite. That is his glory. Kierkegaard once said religion deals with what concerns man unconditionally. In contrast (but with equal exaggeration) one may say that mathematics deals with the things which are of no concern to man at all. Mathematics has the inhuman quality of starlight, brilliant and sharp, but cold. But it seems an irony of creation that man's mind knows how to handle things the better the farther removed they are from the center of his existence. Thus we are cleverest where knowledge matters least: in mathematics, especially in number theory.

Here a passage from Aristotle's *Metaphysics* (982b) comes to my mind where he ponders whether man should ever try to transgress the boundaries of such knowledge as is of immediate concern to him (*tēn kath' heauton epistēmēn*). "Indeed," he continues, "if the poets are right and the Deity is by nature jealous, it is probable that in this case God would be particularly jealous and all those who step beyond (*pantas tous perittous*) are liable to misfortune." I am not so sure whether we mathematicians during the last decades have not "stepped beyond" the human realm by our abstractions. Aristotle (who actually speaks about Metaphysics rather than Mathematics) comforts us by hinting that the envy of the Gods is but a lie of the poets (for "poets tell many a lie," as the proverb says). For us today the idea that the Gods from which we wrestled the secret of knowledge by symbolic construction will revenge our *hubris* has taken on a quite concrete form. For who can close his eyes against the menace of our own self-destruction by science? The alarming fact is that the rapid progress of scientific knowledge is not paralleled by a corresponding growth of man's moral strength and responsibility, which have hardly changed in historical times. I think it is futile to claim with Hardy for mathematics an exceptional and relatively innocent position in this regard. He maintains that mathematics is a useless science, and this means, he says, that it can contribute directly neither to the exploitation of our fellow-men nor to their extermination. However, the power of science rests on the combination of experiment, i.e., observation under freely chosen conditions, with *symbolic construction*, and the latter is its mathematical aspect. Thus if science is found guilty, mathematics cannot evade the verdict.

The simplest, and in a certain sense most profound, example of symbolic construction are the natural numbers $1, 2, 3, \ldots$ by which we count objects. The most natural symbols for them are strokes, one put after the other, $|, \; ||, \; |||,$ $\ldots$. The objects may disperse, "melt, thaw and resolve themselves into a dew," but we thus can keep the record of their number.[1] What is more, we can by a constructive process decide for two numbers represented through symbols which one is the larger, namely by checking one symbol against the other, stroke by stroke. This process reveals differences not manifest in direct observation, which in most instances is incapable of distinguishing even between such low numbers as 21 and 22. We are so familiar with these miracles which the number symbols perform that we no longer wonder at them. But this is only the prelude to the mathematical step proper. We do not leave it to chance which numbers we actually meet by counting this or that concrete set of objects, but we generate the open sequence of *all possible* numbers which starts with 1 (or 0 = nothing) and proceeds by adding to any number symbol n already reached one more stroke, whereby it changes into the following number n'. *Being* is thus projected onto the background of the *possible*, more precisely onto a manifold of possibilities which unfolds by iterating the same step again and again and remains open into infinity. Whatever number n we are given, we always deem it possible to pass to the next n'. This intuition of the "ever one more," of the open countable infinity, is basic for all mathematics; it provides the simplest example of what I like to call an a priori surveyable range of variability. Other examples will come into ken later. At the moment I wanted only to conjure up in preliminary fashion the idea of symbolic construction.

With the constructive there vies the *axiomatic* approach, that has come to play an enormously increased role in our twentieth-century mathematics — to such an extent that in an article by the famous Nicolas Bourbaki on "The Architecture of Mathematics" [1950], this aspect of mathematics dominates the entire picture. Even though I do not quite agree with this attitude I shall now first try to give you a description of mathematical axiomatics. In doing so I follow partly a paper of mine, "A Half-Century of Mathematics,"[2] and also will use notes from the Introduction of a course on "Methods and Problems of Twentieth Century Mathematics," which I boldly began in Zurich and then in Princeton during the years 1952–1953 but stopped before having exhausted even one-tenth of the material.

Whereas the axiomatic method was formerly used merely for the purpose of elucidating the foundations on which we build, it has now become a tool for concrete mathematical research. It is perhaps in algebra that it has scored its greatest successes. Take for instance the system of real numbers. It is like a Janus head facing in two directions; on the one side it is the field of the algebraic operations of addition and multiplication; on the other hand it is a continuous manifold, the parts of which are connected as to defy exact isolation from each other. The one is the algebraic, the other the topological face of numbers. Modern axiomatics, simple-minded as it is, does not like (in contrast to modern politics) such ambiguous mixtures of peace and war, and therefore cleanly separated both aspects from each other.

Let us look for a moment at the algebraic face. A system of elements in which the binary operations of addition and multiplication are defined and satisfy the usual algebraic axioms — I must here forego to enumerate them — is called a *field*. The real numbers form a field, but also the more elementary set of rational numbers. Over the field of rational numbers one can erect the algebraic number fields; if ϑ is the root of an irreducible algebraic equation with rational coefficients the polynomials of ϑ with rational coefficients form such a field. The wonderful and profound theory of algebraic number fields deals with them. While in the old axiomatics one was mostly concerned with axioms which determine the structure of the system completely, as, e.g., the axioms of Euclidean geometry do for Euclidean space, we have here, in algebra, to do with axioms satisfied by many different individual number fields that are not mutually isomorphic. The axioms are now not used as in the investigation of foundations for the unique characterization of one all-encompassing structure, but as a common basis for the investigation of individual entities arising by specified constructions.

You all know that the first consistent axiomatic system ever developed was that of Euclidean geometry. In its original meaning, it deals with the real space of the physical universe, of which our mind seems to have adequate intuition. Euclid took the axioms as evident statements about points, lines, and planes in real space. Of these geometric objects he tries to give a descriptive definition, but all the theorems are developed by rigorous deduction from the axioms and are in no way dependent on the descriptions. A modern form of this sort of geometric axiomatics is Hilbert's *Grundlagen der Geometrie* [*Foundations of Geometry*, Hilbert 1971] published just at the turn of the century. Of course there are obvious improvements. Euclid's list of axioms was still far from being complete; Hilbert's list is complete, and there are no gaps in the deductions. More fundamental is the changed standpoint: Hilbert makes no attempt to describe what points, lines, and planes mean in our spatial intuition; all that we need to know about them and their relations of incidence, congruence, etc., is contained in the axioms. The axioms are, as it were, their implicit (though necessarily incomplete) definitions. Blumenthal records in his life of Hilbert that as early as 1891 Hilbert, discussing a paper by H. Wiener, made a remark which expressed this standpoint in typically Hilbertian manner: "It must be possible to replace in all geometric statements the words point, line, plane by table, chair, mug." In the deductive system of geometry the evidence, even the truth of axioms, is irrelevant; they figure merely as hypotheses of which one sets out to develop the logical consequences. There may exist other material interpretations besides the familiar geometric one for which the axioms turn into true statements; then all the theorems will hold likewise for this interpretation. For instance, the axioms of n-dimensional Euclidean vector geometry hold if a distribution of direct current in a given electric circuit, the n branches of which connect in certain branch points is called a *vector*, and Joule's heat produced in unit time by the current is considered the square of the vector's *length*. The geometric theorem that a vector determines uniquely its perpendicular projection onto a given linear submanifold of the vector space then changes into the statement that a given distribution of electromotive forces in the circuit uniquely determines the current.

From the system of geometry itself, erected on the basis of a few undefinite concepts and a few axioms into which they enter, Hilbert soon passes to a higher level of questions, which one may call the metageometric level; here one inquires into the logical structure of the edifice of geometry, asking in particular for the consistency, the mutual independence and the completeness or categoricity of the axioms. Here his method consists in the construction of models: a certain algebraically constructed model is shown to satisfy all axioms except one; hence the one that is not satisfied cannot be a consequence of the others. In all this, although the execution shows the master hand, Hilbert is not unique. He had predecessors, above all in Italy and Germany. One outstanding example of the method of models had been known for a long time: the Cayley-Klein model of non-Euclidean geometry. But perhaps it is true that Hilbert was the first who moved with freedom and sovereignty on this meta-geometric level.

Let us now return to comparing this transcendental use of axiomatics to the immanent use as exercised for instance in concrete algebraic research. Euclid's axioms, as he understood them, referred to an object, space, which has an extra-mathematical origin, whereby I mean that space is not, like number, mind's free creation. For one who does not accept their evidence they become hypotheses. In contrast, immanent axiomatics as applied in algebra or topology or some other concrete branch of mathematics is neither based on external evidence nor on hypotheses, but the axioms are proved to hold for the mathematical objects in the individual situations to which the axioms are applied. Thus in algebra and number theory one comes again and again across "ideal divisors," of which one proves that they satisfy this or that set of axioms for ideals. It is useful to have these axioms and their consequences before one's eyes independently of the several contexts in which the notion of divisor turns up in algebraic and arithmetical investigations. Transcendental axiomatics has been successfully applied not only to geometry, but also to such fields of knowledge as mechanics, in particular statics of rigid bodies (Archimedes, Galileo, Huygens), special and general relativity theory, blackbody radiation (Hilbert), biology (J. H. Woodger) and economics (von Neumann's and Morgenstern's "Theory of Games"). For the beginner, if I may judge by my own experience as a young student, transcendental axiomatics has something paradoxical and shocking, because one must try to learn to abstract radically from the familiar intuitive significance of the terms occurring in the axioms as undefined concepts. Axiomatics as used in concrete mathematical research is *au fond* a much simpler affair. One just formulates some of the fundamental facts which, as one can prove, hold for the object of investigation, which is usually a set of freely constructed elements. The questions of consistency, independence and completeness are here but of minor importance.

The purpose of this sort of axiomatics is to *understand*; and what it reveals is a limited number of interrelated *structures*, that seem to form the backbone of the mathematical universe. This hierarchy of structures is not closed, but still in the process of growth and unification. We are not content to get convicted, as it were, rather than convinced of a mathematical truth by a long chain of formal inferences and calculations leading us blindfolded from link to link. We would like to be shown not only the goal but also the way and general outline we are to travel to that goal, to understand the underlying ideas of the proof and their connections. Indeed a modern mathematical proof, just as any modern machine or experimental arrangement effaces, so to speak, by the complexity of technical

details, the simple principles on which it is based. When in his lectures on the
history of mathematics in the nineteenth century Felix Klein tries to Riemann
his place he says:

> Certainly the keystone of the edifice of any mathematical theory are
> the stringent proofs of its propositions. By renouncing these, mathe-
> matics would pass its own sentence. Yet it will be forever the secret
> of a genius's productivity to ferret out new problems and to define
> new and unexpected results and connections. Without the disclosure
> of new viewpoints and goals mathematics would soon exhaust itself
> in its pursuit of strict logical reasoning and begin to stagnate for lack
> of new material. Thus in a way mathematics has been advanced most
> by those whose strength lay in intuition rather than logical rigor.

The chief organ of Klein's own productivity was this intuitive perception of inter-
connections and relations between separate fields; he failed where concentrated
force was required. In his memorial address on Lejeune Dirichlet, Minkowski
confronted that minimal principle, first formulated and put to manifold use by
William Thomson (Lord Kelvin) but since Riemann attached to Dirichlet's name,
with what he called the true Dirichlet principle: namely to solve the problems by
a minimum of blind calculation, a maximum of seeing ideas. From this principle,
he said, dates the new age in the history of mathematics.[3]

Ideas and intuitive understanding versus calculation and strict logical deductions: what is the real point in these contrasts, what is the secret of understanding a mathematical fact? Certain epistemological schools, I mention the name
of Wilhelm Dilthey, have claimed understanding from within, hermeneutics, as
the proper basis for historical research and the humanities, whereas the natural
sciences seek to explain rather than to understand the phenomena. The words
"intuition, understanding" appear here with a certain nimbus indicating a depth
and immediacy of their own. In mathematics we prefer to look at things more
soberly. I shall not try, and it would admittedly be difficult, to give a sufficiently
precise description of the mental acts here involved. But I would like to point
out at least one decisive characteristic of the process of understanding.

In order to understand a complex mathematical situation we *separate* in a
natural manner the various sides of the subject in question, make each side accessible by a relatively narrow and easily surveyable group of notions and facts
formulated in terms of these notions, and finally return to the whole by uniting
the partial results in their proper specialization. The last synthetic act is purely
mechanical. The art lies in the first, the analytic act of suitable separation and
generalization. Our mathematics of the last decades has wallowed in generalizations and formalizations. But one misunderstands this tendency if one thinks
that generality was sought merely for generality's sake. The real aim is simplicity: natural generalization simplifies since it reduces the assumptions that have
to be taken into account. Of course, it may happen that different directions of

generalizations make us understand the specific concrete situation under different aspects. Then it is dogmatic and arbitrary to speak of *the true* reason, *the true* source of the particular fact. It is not easy to say what constitutes a *natural* separation and generalization. For this there is ultimately no other criterion but fruitfulness: the success decides. In following this procedure the individual investigator is guided by more or less obvious analogies and by instinctive discernment of the essential, acquired through accumulated research experience. When systematized, the procedure leads straight to axiomatics. Since the dissolution of complex and relatively concrete notions and facts into simple but general and abstract notions attracted so much of the mathematicians' attention in recent times, it is but human that they sometimes indulged in cheap generalizations, generalizations by rarefaction as Pólya called them, which do not add to the mathematical substance but only dilute the good nourishing soup with water. But such degenerations do 'not speak against the basic soundness and importance of the axiomatic approach.

I spent my mathematical youth in Göttingen, Germany, before the First World War, in the era of Felix Klein—David Hilbert. It is interesting to compare the attitude of these two men towards axiomatics. In the time of Klein's productivity (which had passed when I entered the University of Göttingen in 1904) the intuitive realization of inner connections between various domains had been the most characteristic feature of his achievements. Typical is his book on the Icosahedron in which geometry, algebra, function- and group-theory blend in polyphonic harmony.[4] But by instinct he shrank from the process of isolating the components. Therefore he disliked the systematization of his way of understanding things in the form of a rigorous axiomatics; even in analysing he did not want for a moment to lose sight of the whole. Only once, when he had to defend his position concerning non-Euclidean geometry by clarifying explication, was he driven to the bitter end of axiomatization. On the whole his way of thinking prevented him from doing justice to axiomatic research. I remember that once in conversation with me he said: "Suppose I have solved a problem; I have taken a hurdle or jumped a ditch. Then you axiomaticians come around and ask: Can you still do it after tying a chair to your leg?"

How different was Hilbert! He was the champion of axiomatics. I mentioned his *Grundlagen der Geometrie*. But as a matter of fact axiomatics was for him a universal, nay *the* central methods of scientific thought. In a lecture given in Zurich in 1917 he said: "Whatever can become an object for scientific thought, will come, as soon as it is ripe for theory, under the rule of the axiomatic method and thus indirectly of mathematics." One is reminded of Kant's famous saying: "In every branch of knowledge there is as much as true science as there is mathematics in it."[5] Hilbert went to the very end of axiomatics by his attempt to save classical mathematics, after the set-theoretic antinomies had shaken its foundation, by a process of consistent formalization which turns the meaningful mathematical propositions into meaningless formulas. The game of deduction played with these formulas proceeds from a small number of formulas serving as axioms and is played according to certain axiomatic rules. It is no longer the truth of the statements symbolized by the formulas that is in question, but merely the consistency of the game of formulas. Hilbert was interested mainly in axiomatics as far as it concerns the foundations or extra-mathematical branches of knowledge. But after the First World War the immanent axiomatic approach to algebra was developed at his side by Emmy Noether and her school. There were partly antecedent, partly parallel developments in this country, while in France the splendid group of young mathematicians guided by the synthetic spirit known under the name of Nicolas Bourbaki carried the work on in a particularly systematic fashion.

It is time now to mention at least a few of the simplest structures defined in terms of axioms. I take my examples from algebra. The concept of a *field* defined by the axioms for the two binary operations of addition and multiplication has been mentioned before. If one suppresses the axiom that every element $a \neq 0$ has an inverse a^{-1} such that aa^{-1} is the unit element e one arrives at the more general notion of a *ring*. The rational numbers form a field ρ, the rational *integers* a ring P. The importance of the notion of a ring lies in the very fact that it provides the axiomatic basis for the arithmetic of integers. The polynomials

$$f(x) = a_0 + a_1 x + \cdots + a_\nu x^\nu$$

of a variable or indeterminate x with arbitrary rational coefficients a_i form a ring, if addition and multiplication are formally defined in the usual manner; but they do not form a field. Quite generally, given any ring P, the polynomials of x with arbitrary coefficients in P form a ring $P[x]$, which is thus derived from P by a simple algebraic construction. Even more fundamental than the concepts of field and ring is that of a group.

The rotations of a sphere form a group of one-to-one mappings of this sphere into itself, because with any two rotations, S carrying the arbitrary point p into p', and T carrying p' into p'', also the composite transformation ST, carrying p into p'' and obtained by performing first S then T, is a rotation. As the example shows, this composition of transformations or one-to-one mappings of a given manifold into itself is not necessarily commutative, $ST = TS$, but it certainly satisfies the associative law $S_1(S_2 S_3) = (S_1 S_2)S_3$. The identity E carrying every point p of the manifold into itself is a transformation, and with S, mapping p into p', also the inverse S^{-1}, sending p' back to p, is a transformation. A group Γ of transformations is characterized by these three requirements: E belongs to Γ; with any transformation S its inverse S^{-1} is in Γ; with any two transformations S, T also the composite ST is in Γ. Let point configurations on a given manifold be called equivalent, provided one is carried into the other by a transformation S of the given group Γ. Then these axioms correspond to the statements: Every configuration F is equivalent to itself; if a configuration F is equivalent to F', then F' is equivalent to F; if F is equivalent to F' and F' is equivalent to F''', then F is equivalent to F''.

It has proved quite important to emphasize the mere structure of a transformation group by ignoring the nature of its elements and paying attention to the fact only that two elements s, t of the group give rise by "composition" to an element st. An abstract group is given once this binary operation for its elements is defined in such manner that the associative law $s_1(s_2s_3) = (s_1s_2)s_3$ is fulfilled for any group elements s_1, s_2, s_3. Two supplementary axioms are added: (1) There is a unit element e such that $es = se = s$ for every group element s; (2) every group element s has an inverse s^{-1} such that $ss^{-1} = s^{-1}s = e$. It is most astonishing how fertile in consequences these simple assumptions about composition of group elements prove.

A ring is a group with respect to addition (i.e. if addition is taken as the operation of composition by which two group elements give rise to a composite element); the elements of a field which are $\neq 0$ form a group with respect to multiplication.

This quick glance at the axiomatic procedure is perhaps sufficient to illustrate one important fact, namely that it does not lead to lots of small splinters of axiomatically defined structures, but only to a few of marked singular character, like field, ring, ideal, group, vector space. These structures form a sort of hierarchy (e.g. all fields are rings, ideals in a ring are special sub-rings, rings and vector spaces are commutative groups with respect to addition, etc.) and they are also capable of combination (e.g. an axiomatic topology may be imposed on a group, so that we obtain the "double structure": topological group.)

I said that one makes a complex mathematical situation manageable by separation into parts each surveyable by a relatively simple set of axioms. This is one side of the axiomatic procedure. The other side is economy through unification. Again and again it happens that theories which in their intrinsic proper meanings have nothing to do with each other prove to be ruled by the same axioms after a suitable translation of the basic terms of the one domain into those of the other. By this and by the often astonishing simplification of proofs which the axiomatic methods brought about, mathematics has become more unified — in spite of the ever increasing diversity of its problems. Without such unification it would become. humanly impossible to further advance the boundaries of our science. For the human brain has but limited capacity. This tendency of several branches of mathematics to coalesce is another conspicuous feature of its modern development.

Monsieur Bourbaki in the article mentioned before has this to say about it: "The internal evolution of mathematical science has, in spite of appearance, brought about a closer unity among its different parts, so as to create something like a central nucleus that is more coherent than it has ever been." And then a little later he portrays mathematics as follows: "It is like a big city, whose outlying districts and suburbs encroach incessantly and in a somewhat chaotic manner on the surrounding country, while the center is rebuilt from time to time, each time in accordance with a more clearly conceived plan and a more majestic order, tearing down the old sections with their labyrinths of alleys and projecting towards the periphery new avenues more direct, broader, and more commodious."

As I confessed before, this description seems to me to exaggerate somewhat the influence of axiomatics on the architecture and development of our science. And it is high time for me now to confront and combine with the axiomatic the genetic or constructive approach, and thus come to the theme proper of this lecture. Otherwise a lopsided picture would result.

The genetic standpoint seems the natural one for numbers; for they are symbols created for recording. The genesis of the natural numbers $1, 2, 3, \ldots$ has been briefly described at the beginning. From there simple constructive processes lead to the integers which include zero and the negative besides the positive integers, and from them to the rational numbers, with which one reaches for the first time the closure of a field. Maybe algebra would have been inclined to stop here. But the Greeks discovered that the rational numbers are not sufficiently wide to describe all exact measurements in science. This requires the larger field of real numbers. Constructive transition to the continuum of all real numbers is a much more serious affair than the previous steps, and I am bold enough to say that not even to this day are the logical issues involved in that constructive concept completely clarified and settled. The sequence of natural numbers and the continuum of real numbers are the most important examples of ranges of variability that are, in a certain sense, a priori surveyable because they are the mind's own free creations. But the infinitude toward which they open is not that of the given, but of the possible. Wherever mathematics is applied to science (including mathematics itself) we encounter these features: 1) variables whose possible values belong to a range of possibilities on which we have a sure grip because it sprang from our own free symbolic construction; and 2) functions or freely constructed mappings of the range of one such variable upon that of another. The importance of topology rests on the fact that it tries to give mathematical expression to these basic features in the most general way.

In a fashion similar to the natural numbers, our conception of *space* is dependent on a constructive grasp on all possible places. Let us consider a metallic disk in a plane E. Places on the disk can be marked *in concreto* by scratching little crosses on the plate. But relatively to two axes of coordinates and a standard length scratched into the plate we can also put ideal marks in the plane outside the disk by giving the numerical values of their two coordinates. Each coordinate varies over the a priori constructed range of real numbers. In this way astronomy uses our solid earth as a base for plumbing the sidereal spaces. What a marvelous feat of imagination when the Greeks first constructed the shadows which earth and moon, illumined by the sun, cast in empty space and thus explained the eclipses of sun and moon!

The same method of a priori construction is used to subject all phenomena of nature to quantitative analysis. I believe the word *quantitative*, if one can give it a meaning at all, ought to be interpreted in this wide sense. As material of the a priori construction, Galileo and Newton used certain features of reality like space and time, which they considered as objective, in opposition to the subjective sense qualities, which they discarded. Hence the important role played by geometric figures in their physics. You probably know Galileo's words in the *Saggiatore* [*Assayer*] where he says that no one can read the great book of nature "unless he has mastered the code in which it is composed, that is, the geometrical figures and the necessary relations between them." Later we have learned that none of these features distilled from our immediate observation, not even space and time, have a right to survive in a pretended truly objective world, and thus have gradually and ultimately come to adopt a symbolic combinatorial construction.

But I want to come back to the axiomatics of algebra to show how constructive processes in algebra intermingle with the axiomatic foundations. Of an arbitrary element a in a field (or a ring) one forms the multiples

$$1a = a, \ 2a = 1a + a, \ 3a = 2a + a, \ldots$$

and thus νa for an arbitrary "multiplier" ν; this multiplier is clearly not an element of the field but a natural number arising from the constructive process which creates the sequence $1, 2, 3, \ldots$ into infinity. The field has a (unique) unit element e such that $ae = ea = a$ for every element a. There may be a multiplier ν such that $\nu e = 0$. Then also $\nu a = \nu(ea) = (\nu e)a = 0$ for every element a. The least such ν must be a prime. Indeed a factorization $\nu = \nu_1 \nu_2$ would lead through $\nu e = (\nu_1 e) \cdot (\nu_2 e) = 0$ to one of the equations $\nu_1 e = 0$ or $\nu_2 e = 0$. It is therefore impossible that both ν_i are less than ν, hence one must be equal to ν and the other equal to 1. Consequently, either *no* multiple νe of the unit e is equal to zero (field of characteristic ∞), or there is a prime π such that $\pi e = 0$, and then $\pi a = 0$ for every element a (field of prime characteristic π). This shows how the constructed number $1, 2, 3, \ldots$ and their arithmetics enter into the realm of the "axiomatic" numbers a.

Another example of a construction in algebra, already mentioned before, is the adjunction of an indeterminate x: Given an arbitrary ring P of elements, the polynomials of the indeterminate x with coefficients from P form a new ring $P[x]$. An "ideal" in a ring is a subset of the ring elements of such kind that the difference of two elements of the ideal again lies in the ideal and also the product of an element of the ideal by any element of the ring. This notion of ideal is the basis for the arithmetical theory of divisibility. Now here is a further constructive process of great importance which transmutes one ring P into another: Let r be an ideal in P and identify any two elements r_1, r_2 in P which are congruent modulo r, i.e., whose difference $r_2 - r_1$ lies in r. This process of identification carries P into a ring P_r, the "ring of residue classes modulo r." For instance let P be the ring of integers and let r consist of all integral multiples of a prime number p. Our process carries the ring of integers over into one consisting of p elements only which can be represented by the residues $0, 1, \ldots, p-1$ the integers leave when divided by p. Intuitively one may describe this process by rolling a straight line with equidistant marks of distance 1 (the integers) upon a circle of circumference p. It can be shown that this finite ring just because p is prime, is even a field. Indeed for any integer a not divisible by p there exists an integer a' such that aa' is congruent 1 mod p.

Topology provides no less interesting illustrations for the the way in which constructive processes penetrate into axiomatics. But my time is running out and therefore these few algebraic examples must suffice to drive home the point I wanted to make in this lecture: *that large parts of modern mathematical research are based on a dexterous blending of constructive and axiomatic procedures.* One should be content to note their mutual interlocking. But temptation is great, and not all authors have resisted it, to adopt one of these two views as the genuine primordial way of mathematical thinking to which the other merely plays a subservient role. In an address at the Bi-Centennial Conference of the University of Pennsylvania on a similar subject in 1940, I used the foundations of combinatorial topology, instead of algebra, for illustrations and let the strongest light fall on the constructive processes.[6] Indeed my own heart draws me to the side of constructivism. Thus it cost me some effort today to follow the opposite direction, putting axiomatics before construction, but justice seemed to require this from me.

Notes

[This is a complete transcription of manuscript Hs 91a:27 in the Weyl *Nachlaß* of the ETH Archiv in Zürich, previously unpublished in this form. An edited and abridged version was previously published as Weyl 1985, with a helpful introduction by Tito Tonietti. The present transcription includes Weyl's handwritten corrections to the typescript and restores his original paragraphing, as well as including a passage omitted in Weyl 1985 and correcting a few misreadings of his handwriting. Between the two complete typed copies preserved in this manuscript, there are only a very few slight discrepancies; I have followed what seemed the later (and final) copy, which appears first in Hs 91a:27.]

[1] [Shakespeare, *Hamlet* I.ii.129–130.]

[2] [See above, 159–190.]

[3] [For Dirichlet's principle, see above 180.]

[4] [See Klein 1956.]

[5] [At this point in his typescript, Weyl struck out the following sentence: "In his lectures, Hilbert liked to elucidate the method by examples taken from mechanics, genetics, economics."]

[6] [See above, 75–80.]

Why is the World Four-Dimensional?
(1955)

The title of my lecture seems to promise an answer to the question: Why has the world exactly four and not any other number of dimensions? If you have come here to learn that answer, I am bound to disappoint you. My aim is more modest: I shall try to clarify the question somewhat and indicate possible directions in which the answer may lie.

To begin with, let us rather talk about *space* than about the world. *Space is 3-dimensional.* A line is 1-dimensional, a plane 2-dimensional, and space 3-dimensional. There are two interpretations of these statements, one in terms of *elementary geometry* or *special relativity*, the other in terms of infinitesimal geometry or *general relativity*. I find it a little easier to discuss the notion of dimension by means of the 2-dimensional *plane* instead of 3-dimensional space. With respect to a Cartesian coordinate system consisting of an origin O and

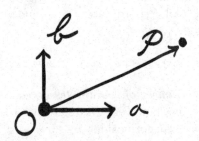

two perpendicular vectors a, b of length 1, each point P may be characterized by its pair of coordinates $(\xi, \eta), \overrightarrow{OP} = \xi a + \eta b$. The points P of the plane are mapped in a one-to-one continuous fashion upon the pairs of real numbers (ξ, η). The coordinates (ξ', η') of the same arbitrary point P with respect to any other Cartesian coordinate system $O'; a', b'$ are connected with (ξ, η) by a *linear transformation*,

$$\xi' = \alpha + a_{11}\xi + a_{12}\eta, \qquad \eta' = \beta + a_{21}\xi + a_{22}\eta,$$

the constant coefficients of which satisfy certain relations ("orthogonal transformations"). *Objective geometric relations* between points are reflected in such arithmetical relations between their coordinates as are *invariant* under these orthogonal transformations, which describe the transition from one Cartesian coordinate system to another. This is the elementary viewpoint in analytic geometry and the principle of special relativity.

But there are many ways of introducing coordinates in the plane; e.g. for some purposes the polar coordinates r, φ are more advantageous. Let us call a pair of numbers (x_1, x_2) an *arithmetical point* and the totality of all such points the arithmetical plane. Coordinatization then consists in a one-to-one continuous *mapping* of the geometric plane upon the arithmetical plane (the mapping, though, may be restricted to portions of the two planes). Saying that the plane is 2-dimensional means that we need two numbers for coordinatization. Surfaces are two-dimensional manifolds, although on them we have no such special coordinates as the Cartesian ones for the plane. E.g. on the globe we may fix the position of a ship by latitude ϑ and longitude φ. In interpreting ϑ and φ as Cartesian coordinates in a Euclidean plane one obtains a *plane map of the globe* (or of part of the globe). Vice versa, any plane map of the globe stands essentially for a definite coordinatization of it. The possible positions of a "top," a rigid body that can freely rotate around a fixed point O, form a 3-dimensional manifold: the top has 3 degrees of freedom, while the possible positions of a freely mobile rigid body constitute a variety of 6 dimensions. The *spectral colors* form a 1-dimensional manifold.[1] The possible states of a *gas* form a 2-dimensional continuum, because such a state is to be specified by two independent quantities, e.g. pressure and temperature. It is the gist of general relativity that it admits, on an equal footing as it were, every possible coordinatization.

Transition from one coordinate system to another is effected by transformation. Thus Cartesian and polar coordinates, (ξ, η) and (r, φ), are related by the transformation formulas

$$\xi = r \cos \varphi, \qquad \eta = r \sin \varphi.$$

In order to show that the *number of coordinates* necessary to cover a given continuous manifold is the same, however coordinatization is carried out, one must prove that two continuous transformations $x \to y, y \to x$,

(1)
$$
\begin{aligned}
y_k = \psi_k(x_1, \ldots, x_m) & \qquad\qquad x_i = \varphi(y_1, \ldots, y_n) \\
(k = 1, \ldots, n) & \qquad\qquad (i = 1, \ldots, m)
\end{aligned}
$$

mediating between m variables x and n variables y cannot be inverse to each other without having $m = n$. That the two transformations (1) defined by the n functions ψ_k of the m variables x and the m functions φ_i of the n variables y are inverse to each other means that by carrying out the two transformations in succession, in either of the two possible orders, one obtains the identity:

$$\varphi_i(\psi_1(x), \ldots, \psi_n(x)) = x_i \quad | \quad \psi_k(\varphi_1(y), \ldots, \varphi_m(y)) = y_k.$$

If we limit ourselves to *differentiable transformations* — and present-day physics has no reason to go beyond that — the equation $m = n$, which guarantees the invariance of dimensionality, can be proved by means of the *general idea underlying differential calculus*, namely the fact that in the infinitesimally small all relations become linear. Development of classical physics supported the conviction that the truly elementary laws of of nature are laws of near-by action, connecting the values of state quantities but at infinitely close points in space and infinitely near moments in time. At a fixed "point"

$$x_1^\circ, \ldots, x_m^\circ \quad | \quad y_1^\circ, \ldots, y_n^\circ,$$

we obtain for corresponding infinitesimal increments dx and dy of the coordinates x and y from (1) such linear equations as

$$dy_k = \sum_i \beta_{ki} dx_i, \qquad dx_i = \sum_k \alpha_{ik} dy_k.$$

They express two mutually inverse linear substitutions of the variables dx_i and dy_k. From the theorem that a system of homogeneous linear equations

$$\sum_k \alpha_{ik} \eta_k = 0$$

always has a non-trivial solution η, provided the number of equations is less than that of the unknowns, there follows easily that the existence of such mutually inverse linear transformations is neither compatible with $m < n$ nor with $n < m$, hence requires $m = n$.

The metrical structure of *Euclidean space* makes it possible to exhibit a special kind of coordinates named after Descartes, such that the transition from one of these distinguished coordinate systems to another is expressed by a linear transformation. By *Einstein's theory of gravitation* we have learned that the actual physical space is not of this special structure (although the deviations are small). But the physical space is at least such that it distinguishes the "differentiable coordinate systems." Any two of these are connected by transformation functions φ, ψ, cf. (1), which are not only continuous but have continuous first derivatives. For otherwise such simple and indispensable physical notions as that of velocity would be without foundation. Indeed this notion requires differentiation of the positional coordinates x_i of a moving mass point with respect to the time parameter t. Hence it is sufficient for physics to have the invariance of dimensionality established, as just has been done, for differentiable transformations (1).

Coordinates are nothing but *names* or *labels* by which we try to distinguish points. For points are all alike and by themselves bear no marks by which to recognize them. Simpler than a continuum of infinitely many points is a finite collection of individual elements. Given 15 elements we could label them by the numbers 1, 2, $\ldots$, 15; but we could also label them by the pairs (ξ, η) where ξ ranges from 1 to 5, η from 1 to 3. *One* numerical label accomplishes here the same

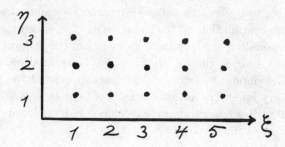

as two simultaneous numerical labels. There is thus no basis for the difference of dimensions. Why then in a continuum? The fact that one can distinguish between a one- and a two-dimensional continuum is obviously due to the very demand of *continuity*, to the requirement that each coordinate be a *continuous* function of the variable element P of the continuum.

I hope enough has now been said to explain the meaning of the statement: Space, i.e. the continuum of space points, is 3-dimensional. The *metrical structure of Euclidean space* can then be expressed as an infinitesimal law as follows: In terms of certain distinguished coordinates, the Cartesian ones, ξ, η, ζ, the square ds^2 of the distance of two infinitely near points P, P' with coordinates ξ, η, ζ, and $\xi + d\xi, \eta + d\eta, \zeta + d\zeta$, respectively, is given by

$$ds^2 = d\xi^2 + d\eta^2 + d\zeta^2 \qquad \text{(Pythagoras)}.$$

Again let us consider the plane instead of the space and thus cancel the third coordinate ζ. The use of infinitesimal quantities like the differentials $d\xi, d\eta$ is horrifying to the modern mathematician. But it is easy to replace them by *differential quotients* to which nobody objects. Imagine a mobile point whose position (ξ, η) in the plane at every moment t is given by $\xi = \xi(t), \eta = \eta(t)$. The formula $ds^2 = d\xi^2 + d\eta^2$ means that the length of the curve the point describes during the time interval $a \leq t \leq b$ equals

$$\int_a^b \frac{ds}{dt} \cdot dt = \int_a^b \sqrt{\left(\frac{d\xi}{dt}\right)^2 + \left(\frac{d\eta}{dt}\right)^2} \cdot dt.$$

The special form of the Pythagorean law $ds^2 = d\xi^2 + d\eta^2$ is bound to the special Cartesian coordinates. How does it read in *arbitrary coordinates* x_1, x_2? Suppose they arise from ξ, η by the transformation

$$\xi = \xi(x_1, x_2), \qquad \eta = \eta(x_1, x_2).$$

You have simply to substitute for $d\xi, d\eta$ their expressions

$$d\xi = \frac{\partial \xi}{\partial x_1} dx_1 + \frac{\partial \xi}{\partial x_2} dx_2, \qquad d\eta = \frac{\partial \eta}{\partial x_1} dx_1 + \frac{\partial \eta}{\partial x_2} dx_2$$

and get the quadratic form

$$(2) \qquad ds^2 = g_{11}dx_1^2 + 2g_{12}dx_1dx_2 + g_{22}dx_2^2 = \sum_{i,j=1}^{2} g_{ij}dx_idx_j$$

of the differentials dx_1, dx_2 with symmetric coefficients $g_{ij} = g_{ji}$. For instance, in terms of polar coordinates

$$ds^2 = dr^2 + r^2d\varphi^2.$$

The coefficients g_{ij} depend on the point $P = (x_1, x_2)$. The relative coordinates dx_1, dx_2 of P' with respect to P are the components of a variable *line element* $\overrightarrow{PP'}$ issuing from P, and the quadratic form (2) expresses how ds^2 depends on these line elements that form the vector compass at P.

A *quadratic form*

$$\sum g_{ij}\xi_i\xi_j \qquad (i,j = 1, \ldots, n; g_{ij} = g_{ji})$$

in variables ξ_i is *non-degenerate* if the determinant $|g_{ij}| \neq 0$, a property that is not destroyed by any invertible linear transformation. The form is *positive definite* if $\sum_{i,j} g_{ij}\xi_i\xi_j > 0$ except for $\xi_1 = \cdots = \xi_n = 0$. Such a form is necessarily non-degenerate. The *unit form* $\xi_1^2 + \cdots + \xi_n^2$ is clearly positive-definite. Any non-degenerate form may by a suitable invertible linear transformation brought into the form

$$-\xi_1^2 - \cdots - \xi_r^2 + \xi_{r+1}^2 + \cdots + \xi_n^2.$$

However this is done, the number r of terms with the minus sign, the so-called *inertial index*, is always the same. The positive-definite forms are those of index $r = 0$.

It was Riemann who first investigated n-dimensional manifolds on which every line element $\overrightarrow{PP'}$ has a square length ds^2 that for a given point $P = (x_1, \ldots, x_n)$ is a positive-definite quadratic form $\sum_{i,j} g_{ij}(x)dx_idx_j$ of the components dx_i of the variable line element $(dx_1, \ldots, dx_n)$ issuing from P. Hence we call such a manifold *Riemannian*.[2] E.g. any surface in 3-dimensional Euclidean space is a 2-dimensional Riemannian manifold. (Of course it is understood that ds^2 is determined by the points P, P', and not affected by the choice of the coordinates x_i to which the manifold is referred. Hence after passing to another coordinate system $\bar{x}_i$ by means of the transformation

$$(3) \qquad x_i = x_i(\bar{x}_1, \ldots, \bar{x}_n), \qquad dx_i = \sum_j \frac{\partial x_i}{\partial \bar{x}_j}d\bar{x}_j$$

one has

$$(4) \qquad ds^2 = \sum_{i,j} \bar{g}_{ij}(\bar{x})d\bar{x}_id\bar{x}_j$$

where the right side of (4) is obtained from (2) by the substitution (3).) The Riemannian manifold is called *flat* if coordinates x_i exist in terms of which ds^2 is expressed everywhere by the unit form,

$$ds^2 = dx_1^2 + \cdots + dx_n^2.$$

One can show that in that case any two coordinate systems in terms of which ds^2 has this simple expression are connected by a linear "orthogonal" transformation. The 3-dimensional Euclidean space is completely described as a flat Riemannian manifold of dimensionality 3. Indeed all concepts and facts of Euclidean geometry can be defined in terms of, and can be derived from, this description. *The lawfulness of our space is of such nature that it carries over in an absolutely cogent manner from 3 to any number of dimensions*, simply because the notions of quadratic form, of positive-definite quadratic form and of unit form, are clearly not limited to 3 variables. In modelling actual space in accordance with these simple mathematical notions the Creator showed, if that is not a blasphemy to say, a supreme taste for mathematical harmony. In view of this perfect harmony we are inclined to ascribe a sort of necessity to His construction. But then we must admit that, in comparison to this necessity, the choice of 3 dimensions appears as a completely arbitrary act; any other number of dimensions would have been just as good. Thus to the question: Why has space 3 dimensions, we can at this step merely answer: The harmonious laws describing the structure of space give no reason whatsoever for its having 3 rather than 1 or 129 dimensions.

With respect to the distinction of general Riemannian, and the more special flat, manifolds it ought to be remarked that any *manifold* of lesser dimension *embedded in a Riemannian one* is again Riemannian, whereas a manifold embedded in a flat Riemannian space, e.g. a surface in Euclidean 3-space, is in general not flat but of course Riemannian.

The events in this world happen in space *and time.* Therefore we now have to include time in our discussion. A strictly localized event takes place at a definite point (ξ, η, ζ) of space (for the moment to be assumed Euclidean) and at a definite time τ; or at a definite *space-time-point* $(\xi, \eta, \zeta; \tau)$, "here-now." The totality of space-time-points, of possible localizations in space and time, is what, with Minkowski, we call the *world.* Thus space having 3 and time 1 dimension, the world is four-dimensional. We have here described the space point-by its coordinates ξ, η, ζ in a definite Cartesian system. In the same manner τ measures time in terms of an arbitrarily chosen beginning, e.g. Christ's birth, and a unit, day or hour or minute or second. To make possible a *graphic representation* we deprive space of one of its dimensions and then picture it as a horizontal plane E while we plot time τ on a vertical axis. All events occurring at the same point in space then lie on a vertical; the vertical is the world-line of a mass point that is *at rest.* All events that happen simultaneously lie in a horizontal plane. He who with Newton believes in *absolute space* and *absolute time* will, therefore, ascribe to the world a stratification in horizontal layers and a cross-fibration into vertical fibres. Newton confesses in the Preface of his *Principia* that he wrote his treatise for the express purpose: to determine the *absolute* motion of bodies from their observable *relative* motions in combination with the *forces* acting upon them. (E.g. the tension of a thread joining two balls indicates that they do not rest but rotate around each other.) However Newton did not succeed completely; certain metaphysical beliefs prevented him from realizing his failure. He succeeded only in distinguishing uniform motion (motion in a straight line with constant velocity, represented in our graphical picture by a straight world-line) from all other motions. And we know now quite definitely that one cannot go beyond that. This does away with absolute space. What about absolute time, the layers of simultaneity? Their significance is originally based on the naive assumption that an event takes place at the same moment it is observed. Olaf Römer's discovery of the finite velocity of light knocked that foundation out. Let us represent in our diagram 1 sec and the distance light travels in 1 sec by vertical and horizontal segments of equal lengths [3] Then the world-points at which a light signal given at the world point O, here-now, arrives, will form a vertical circular cone with vertex in 0 and an opening angle of 90°. To this forward light cone we add the backward part. The whole cone consists of the world points $(\xi, \eta, \zeta; \tau)$ satisfying the equation

$$\tau^2 - (\xi^2 + \eta^2 + \zeta^2) = 0$$

(provided the origin O has the coordinates $(0,0,0,0)$). The very meaning of the horizontal layers of simultaneity is that they are supposed to describe the *causal structure* of the world. Any action at O can influence only events taking place in the half of the world above the layer through O (active future of O), while all events which can possibly have influenced that which happens at O must lie in the lower half (passive past of O). But numerous experiences have taught us that no effects travel faster than light. The light cones,

$$(\tau - \tau_0)^2 - (\xi - \xi_0)^2 - (\eta - \eta_0)^2 - (\zeta - \zeta_0)^2 = 0,$$

rather than the horizontal layers, $\tau = \text{const.} = \tau_0$, therefore, determine the causal structure. The active future of O consists of the interior of the forward cone, its passive past of the interior of the backward cone. (They are separated from each other by a world region which is not causally connected with O at all.) In this way Einstein and Minkowski arrived at the conclusion (special relativity theory) that the world is a 4-dimensional flat manifold, with the slight difference, however, over against the classical Euclidean geometry, that the fundamental quadratic form

$$dx_1^2 + dx_2^2 + dx_3^2 - dx_4^2$$

defining the metric is of inertial index 1 and not 0.

While physicists and mathematicians were busy to swallow this revolutionary insight, Einstein pushed further ahead. In a few words I shall try to repeat the argument which forced him to replace special by *general relativity*, namely to replace the flat space-time manifold by a more general Riemannian 4-dimensional manifold of index 1; i.e. by a world the metric of which is expressed in terms of arbitrary coordinates $x_1, \ldots, x_4$ by a quadratic differential form

$$ds^2 = \sum g_{ij} dx_i dx_j \qquad (i, j = 1, 2, 3, 4)$$

of index 1 with non-constant and *a priori unpredictable* coefficients g_{ij}. By calling them "a priori unpredictable" I want to indicate that the g_{ij} are not given once for all, but have to be determined a posteriori by the observation of actual events; they describe a physical field as real as the electromagnetic field, that is acting upon matter but is also reacted upon by matter. — Galileo had taught us to understand the motion of bodies as resulting from a struggle between inertia and external force. The motion of a mass point is represented by a world line in the 4-dimensional world. If not acted upon by any external force, its world line will be a geodesic of the metrical field. Thus the inertial motion is determined by the g_{ij}. Certain expositions of general relativity theory used to stress, as E. Mach had done, the conception that only *relative* motion of bodies counts. Suppose a train collides with another near a village church. The train relative to the church steeple, or the steeple relative to the train, experiences a violent jerk. Why then, asked Lenard, Einstein's passionate adversary among the physicists, is the train smashed up and not the steeple? The answer, I think, is clear: the train goes to pieces because of the violent clash between its inertia and the molecular forces exerted by the engine of the colliding train, while the church tower is left quietly to follow its inertial motion. Thus the metric field is revealed as an agent of formidable reality. Now something that acts upon matter in such forcible fashion has no claim to be enthroned in sphinx-like rigidity above the ever-changing world of matter; it should itself be flexible and respond to the changing distribution of matter; he who makes suffer must suffer himself. And the fact that gravity of a mass point is always strictly proportional to its inertial mass becomes understandable *ex principio* if gravitation can be considered the changing part of inertia; thus in the juxtaposition of inertia and force gravitation belongs to the side of inertia inseparably blended into it, and not of force. This Einsteinian conception of gravitation met with complete success, and thus we have to give up the fixed Euclidean metric of the 4-dimensional world and replace it by a variable Riemannian metric. This is the point to which I had to lead you before I could begin to discuss our central question: The stage on which this frightful spectacle of the world is performed, its extensive space-time medium, is a 4-dimensional manifold endowed with a variable Riemannian metric of index 1. But the notion of Riemannian manifold is not bound to the special dimensionality 4; *as far as this metrical structure goes, any other dimension would do just as well.* Why then 4?

In discussing this question, we can not limit ourselves to the metrical structure of space-time, we have to look at all the laws of the physical phenomena that take place on this stage and ask: Can they also be carried, in the same cogent and unambiguous manner, from 4 to any other number of dimensions, or do they contain something specific for $n = 4$? The answer is: *They are all such as to allow immediate generalization to n dimensions.* Hence in these laws, there is no reason to be found for the Creator's whim to fashion a 4-dimensional world as the scene for our activities. Of course, our present knowledge of the laws of the physical world is incomplete, and one day it might strike a deeper level on which dimensionality ceases to be indifferent, but at the moment this is merely a hope and not a fact.

Yet even if I am unable to present a satisfactory answer to our question, I think I need not stop here and can do something to further illuminate it. Let us begin with those physical laws we know best: *Maxwell's laws of the electromagnetic field in an empty universe.* I shall not write them down explicitly but I wish to mention that their formulation is particularly succinct in terms of the theory of harmonic linear tensors, which in the last two decades has become a central topic for mathematical research. There are in an n-dimensional Riemannian manifold linear tensors of rank $1, 2, \ldots, n - 1$, and two operations d and δ of differentiation resembling the well-known operations of rot and div in ordinarily 3-dimensional vector analysis.[4] d changes a linear tensor of rank p into one of rank $p + 1$ while δ lowers the rank by 1. A linear tensor f is harmonic if both df and δf vanish. Maxwell's laws in empty space may be summarized as stating that the electromagnetic field is a *harmonic* linear tensor f of rank 2, $df = 0, \delta f = 0$.

Again we ask: *Why the rank 2?* Probably the three questions for the reasons of the dimensionality 4 of the universe, the inertial index 1 of its metric and the rank 2 of the electromagnetic field ought to be answered jointly.

A first remark I wish to make is about the difference of *even and odd* dimensionality. For the moment I assume that the world is an n-dimensional flat manifold (which is true in great approximation), of index 1, and I accept without question the fact that the electromagnetic field is described by a linear tensor of rank 2. Merely the dimensionality n is in question. Suppose that a single light, a candle, is burning in the world. Now blow this candle out; what will happen according to the Maxwellian laws? You probably think it will grow dark, pitch dark, in a sphere around the candle which expands with the velocity of light. And you are right — *provided the number n is even,* especially in our world for which $n = 4$. But it would not be so in a world of odd dimensionality. Now here you have a physically interesting difference at least between even and odd dimensions, although it does not single out the dimensionality 4. "And God said, Let there be light: and there was light," so tells the story of Creation in Genesis, Chapter I. If He wished to keep the possibility open for Himself to say "Let there be darkness again" and to accomplish this by blowing out His candles then He had to make the world of even dimensionality.

This example gives some inkling in which direction we must look for the distinctive characters of dimensionality, inertial index, and rank of the electromagnetic field tensor. Let us start with the index r. There is some good reason for the special value $r = 1$ of the index to be found in the *topology of causa-tion*. In a world of index 0 there is no propagation of effect, in a world of index higher than 1 past and future form one connected domain; only if the index is 1, the "interior of the light cone" (consisting of the line elements at P for which $ds^2 < 0$) exists and falls apart into two disconnected regions: passive past and active future. Whether a domain is connected or consists of several disconnected parts is a topological difference. Hence in the case $r = 1$ the world is topolog-ically distinguished by a very simple and decisive feature from all other cases. This seems to me a genuine and fairly satisfactory answer to the question why $r = 1$.

When Dirac formulated his quantum mechanical equations for the electron in the frame of special relativity the type of equations looked so strange, as if they were bound to the number 4 of dimensions. But alas, in his general investigations on Lie groups E. Cartan had long before established the group-theoretical basis for the corresponding formulas in n dimensions, and later on the n-dimensional spinor theory was worked out in algebraic detail by Richard Brauer and the speaker.[5] In the presence of matter, in particular of electrons and positrons, the second system of Maxwell's equations $\delta f = 0$ has to be replaced by $\delta f = s$ where the linear tensor s of electric current is given in terms of the Dirac quantities ψ defining the electron-positron field. Now it is important for the statistical interpretation of quantum physics that that component of s which defines the probability of the presence of an electron in a given volume should be positive, and this is so only if *the rank of s equals the inertial index* r. But with e denoting the rank of the electromagnetic field, the rank of s *is* $e - 1$. Thus quantum mechanics and its statistical interpretation force the relation $e = r + 1$ upon a prospective world-builder. After he made $r = 1$ because of the topology of causation, he has no choice but take $e = 2$. These seem to me sound arguments for the values $r = 1, e = 2$, of inertial index r and of rank e of the electromagnetic field.

Locally, that is to say in the infinitesimal neighborhood of a.world point P, the equation of the light cone

$$(2') \qquad\qquad ds^2 = \sum_{i,j} g_{ij}\, dx_i dx_j = 0$$

describes the world's causal structure. Knowing the *causal structure*, we do not know the g_{ij} themselves but only their ratios; the causal structure is not changed when the g_{ij} defining the metric are replaced by $\lambda \cdot g_{ij}$ with a positive factor of proportionality λ which is an arbitrary function of P. The mathematician speaks here of conformal structure since the ratios of the g_{ij} determine the angles and the ratios of lengths of the various line elements at P. It would not seem unreasonable to assume that the world is endowed, not with a metric but with a conformal or causal structure only, in other words that only lengths of line elements *at the same point* are comparable to each other. Then the laws of nature would not be affected by the modification $g_{ij} \to \lambda \cdot g_{ij}$ of the metrical field. Now it is a fact that the equations, characterizing a harmonic linear tensor, $df = 0, \delta f = 0$, are invariant under this substitution if and only if $n = 2e$, i.e. if the dimensionality n is twice the rank e of the electromagnetic tensor f. If $r = 1, e = 2$ we would thus obtain the desired $n = 4$.

If r were 0, the equations $e = r + 1, n = 2e$, for which we have argued here, would yield $e = 1, n = 2$; i.e. we would deal with a linear tensor of rank 1, in other words a vector, in a 2-dimensional world of index 0. Every Riemannian manifold of dimension 2 and index 0 has the conformal structure of the Euclidean plane. Combine the plane's real Cartesian coordinates x_1, x_2 into the complex variable $z = x_1 + ix_2$. The vector has 2 components u_1, u_2 which we combine to form the function $w = u_1 - iu_2$ of the complex argument $z = x_1 + ix_2$. That the vector is harmonic amounts to saying that w is an analytic function of z. And indeed it is a well-known and highly important fact that an analytic function of z remains an *analytic function* of z' if the transition $z \to z'$ is a conformal mapping. Thus this lowest case in which there is no propagation of effects ($r = 0$) corresponds to the theory of analytic functions of a complex variable, and lines up very neatly with the next higher case $r = 1, e = 2, n = 4$ corresponding to the theory of propagation of light.

This looks quite harmonious, doesn't it? I wish I could close here, but I am afraid that would be thoroughly dishonest. It is true that the laws of the electromagnetic field in empty space presuppose only a conformal structure for the world, they are invariant with respect to the replacement of g_{ij} by $\lambda \cdot g_{ij}$. But when we pass to the non-homogeneous Maxwell equations $\delta f = s$ which describe the *generation of such fields by matter*, or to Einstein's laws of *gravitation even in empty space*, this is no longer so. The field laws, as derived by Maxwell and Einstein from experience, are not conformally invariant, neither for the gravitational field, nor for the electromagnetic field in the presence of electric charge and current. The fact that Maxwell's equations for the electromagnetic field in empty space, in particular the laws for the propagation of light, are conformal-invariant if and only if $n = 4$, is certainly worth pointing out, but it does not lead us very far.

However, there are indications in recent mathematical research that are apt to raise hopes for a more satisfactory answer in the future. So far we have accepted the Riemannian structure of the world without questioning. If one wants a deeper reason for why the world has this metrical structure, I am sure one has to ask why it is the orthogonal group among all groups of homogeneous linear transformations that is characteristic for the local metric of the world. Why is this metric Pythagorean? For a given non-degenerate quadratic form $\sum g_{ij} x_i x_j$ (with constant coefficients $g_{ij} = g_{ji}$) the orthogonal transformations are those linear transformations $x_i = \sum_j a_{ij} \bar{x}_j$ which leave this form invariant: $\sum g_{ij} x_i x_j = \sum g_{ij} \bar{x}_i \bar{x}_j$. Helmholtz succeeded in characterizing the group of orthogonal transformations or rotations by the degree of freedom which arbitrary rotations impart to a rigid body, a top spinning around a fixed point. However, his characterization is valid only for the group as defined by a positive-definite quadratic form. True, the metric of the 3-dimensional Euclidean space is described by such a form, but not the metric of the 4-dimensional world; the latter depends on a form of the inertial index 1. Moreover Helmholtz undertook his analysis to account for the Euclidean nature of the geometry of the physical bodies and fits into the frame of special, not of general relativity. I have advocated another intrinsic characterization of the orthogonal group which applies to forms of arbitrary index and stresses just that property of this group which is decisive for the building-up of Riemann's generally relativistic geometry: the fact that the metric of the space-time continuum uniquely determines the gravitational field. Both characterizations of the orthogonal group, Helmholtz's as well as mine, hold for every number of dimensions; perhaps they make us understand why God imposed a Pythagorean metric upon this world, but give no hint *why* He made it 4-dimensional.[6] From a purely mathematical standpoint, however, there is a vast variety of ways how one can attempt to distinguish by simple intrinsic properties the orthogonal groups from all other groups of linear transformations. Here is one that quite naturally suggests itself. Let there be given a group G of linear transformations in an n-dimensional vector space. Consider an arbitrary direction at the origin O as indicated by a straight line r through O. I call the direction b in which a point on r moves under the influence of any infinitesimal operation of G *conjugate* to r. For the orthogonal group, b consists of all directions perpendicular to r; b and r are thus related by what geometers call a projective involution. Is this property, that conjugation is a projective involution, characteristic for the orthogonal group! The answer is, *yes*, except for the dimensionalities 7 and 8. For either of these two dimensions there exists, besides the orthogonal group itself, one subgroup of it of 7 parameters less enjoying the same property. Now here something that is specific for the dimensionalities 7 and 8 comes into our ken. In large part of mathematics, e.g. in the theory of linear equations, just as in physics, the general laws are indifferent towards the number of dimensions. But there are also deeper problems for which this ceases to be the case.

The above is but an example. To any group of transformations one can ascribe a *structure* which is represented by the corresponding abstract group. The *structure* of the orthogonal group turns out to be quite different for different dimensions n. All orthogonal groups except that of dimension 4 have the structural property described by the word "simple"; I shall not try to define it here.[7] Only the 4-dimensional orthogonal group has a more complex structure, being the direct product of two simple groups. This fact, however, seems of little interest in physics, and if it is true that Nature loves simplicity would provide a reason *against* rather than *for* the dimensionality 4.

The orthogonal group for a positive-definite quadratic form is closely related to the notion of *sphere*. Topology has during the last years revealed elementary and highly significant differences in the behavior of spheres of different dimensions. They concern the continuous mappings $S_n \to V$ of an n-dimensional sphere S_n into a given manifold or variety V. Such a mapping is considered trivial if you can deform it continuously into one that maps S into a single point of V. We discuss the case where V is also a sphere, a sphere S_{n-d} of lower dimension $n-d$ than the first S_n. The question is whether all mappings $S_n \to S_{n-d}$ are trivial or not. A first result is to the effect that the answer depends only on the difference d of dimensions, at least as soon as $n \geq 2(d-1)$. The answer is *yes* for $d = 4$ and 5, it is *no* for $d = 1, 2, 3, 7$; so in the latter cases there are non-trivial mappings. This is also the case if d is of the form $2p - 3$ where p is a prime number. The notion of homotopy group introduced by Witold Hurewicz about 20 years ago puts this problem in sharper relief, and in this sharpened form it stands in the center of today's topological interests; in spite of many surprising individual results, the mystery which shrouds it has not yet been completely pierced. That a simple property of a spherical space should depend on whether its dimension is a prime number — I admit, it is not exactly this that we found, since d is the difference of two dimensions — is certainly a novel and exciting feature in mathematics.

Huge parts of the theory of spheres or of the orthogonal groups are, as we mentioned before, quite insensitive with respect to dimension. But now we have pointed out some examples of a deeper level, of properties that vary in a strange manner from dimension to dimension. Unfortunately, however, those mathematical features which play a role in present-day physics do not belong to this deeper level. Here theoretical physics lags behind mathematics. But is not at least the hope justified that this will not be so forever, that one day physics will discover laws of nature, the mathematical formulation of which takes account of such structural features as are highly sensitive to dimension. If that ever comes to pass, we can expect to understand the specific character of the world's actual dimensionality 4. At the moment we have not yet reached that point, and instead of explaining to you why the world is 4-dimensional I have not done much more than indicated reasons why we are at present unable to give a satisfactory answer to that question.

Notes

[This previously unpublished manuscript is Hs 91a:33 in the Weyl *Nachlaß* of the ETH Archiv in Zürich; it was delivered as a lecture at the National Academy of Sciences, National Research Council, in Washington, D.C. on March 29, 1955. I thank the ETH Archiv for their kind permission to publish this manuscript and give special thanks to Christine Di Bella, Archivist at the Shelby White and Leon Levy Archives Center at the Institute for Advanced Study, Princeton, who found the archival materials that showed the date of this lecture. The transcription printed here includes Weyl's handwritten corrections to the typescript.]

[1] [In 1854, Riemann already used this example of a non-spatial manifold, which ultimately goes back to Helmholtz; see Pesic 2007, 24, 34n4, 50, Pesic 2012.]

[2] [For Riemann's foundational 1854 address with commentary, see Pesic 2007, 23–40; Weyl himself wrote a famous commentary on this lecture, Riemann 1919, reprinted in Riemann 1990, 740–769.]

[3] [Though Weyl does not include it here, the same diagram can be found above, 71.]

[4] [The present-day notation "curl" is more familiar than Weyl's "rot." By his notation $\delta f = 0$ Weyl indicates the covariant divergence of the electromagnetic field tensor $F^{\mu\nu}_{\|\nu} = 0$, in free space; $df = 0$ signifies the corresponding equation for its dual tensor, $^*F^{\mu\nu}_{\|\nu} = 0$ always (a result of the antisymmetrization of this tensor, physically reflecting the absence of magnetic monopoles). Weyl will shortly discuss the more general case $\delta f = s$, or $F^{\mu\nu}_{\|\nu} = s^{\mu}$, where s^{μ} is the charge=current source-vector. See, for example, Adler Bazin and Schiffer 1975, 93–94, 114–116.]

[5] [See Brauer and Weyl 1935. Cartan 1966 discusses the differences between spinor theories in even and odd dimensions. For the relation between Weyl and Cartan's work, see Scholz 2011a.]

[6] [For Helmholtz's arguments, see Pesic 2007, 47–52; for a fine modern presentation, see Adler Bazin and Schiffer 1975, 10–16.]

[7] [Consider a group G with elements g. A subgroup N of G is called *normal* (or *invariant*) if, for every element g of G and every element n of N, gng^{-1} is always an element n' of N; that is, if $gn = n'g$. By definition, a *simple* group has no normal subgroups, other than itself and the identity; roughly speaking, a simple group cannot be "split" into smaller "atoms." See Pesic 2003, 111–130, for a discussion in the context of the solvability of algebraic equations and Galois theory. Weyl here refers to important results about the structure of the n-dimensional orthogonal group he had found in 1925–1926; see *WGA* 2:543ff. Weyl 1997, 137–164, gives his magisterial treatment of the orthogonal group.]

References

Note that the abbreviations *WGA* refers to Weyl, *Gesammelte Abhandlungen* (Weyl 1968), whose pagination will be used for all references to those works. In multivolume works, 2:111 refers to volume 2, page 111; likewise, HGA refers to David Hilbert, *Gesammelte Abhandlungen* (Hilbert 1932–1935). The listing of editions below is not exhaustive; only the original (first) and most recent (English) editions are listed, where appropriate.

Adler, Ronald. Maurice Bazin, and Menahem Schiffer. 1975. *Introduction to General Relativity*, 2nd ed. New York: McGraw-Hill.

Alunni, C., M. Castellana, D. Ria, A. Rossi, eds. 2009. *Albert Einstein et Hermann Weyl, 1955–2005: Questions épistémologiques ouvertes*. Mandurai: Barbieri.

Anaxagoras of Clazomenae. 2007. *Fragments and Testimonia: A Text and Translation with Notes and Essays*, tr. Patricia Curd. Toronto: University of Toronto Press.

Ash, Mitchell G. and Alfons Söllner (eds). 1996. *Forced Migration and Scientific Change: Emigré German-Speaking Scientists and Scholars after 1933*. Cambridge: Cambridge University Press.

Barrow, J. D. 1983. "Dimensionality." *Philosophical Transactions of the Royal Society of London A* 310:337–346.

Bateman, H. 1909. "The Conformal Transformations of a Space of Four Dimensions and their Applications to Geometrical Optics." *Proceedings of the London Mathematical Society* 7:70–89.

——————. 1910a. "The Transformation of the Electrodynamical Equations." *Proceedings of the London Mathematical Society* 8:223–264.

——————. 1910b. "The Transformations of Coordinates Which Can Be Used to Transform One Physical Problem into Another." *Proceedings of the London Mathematical Society* 8:469–488.

Bell, J. L. 2000. "Hermann Weyl on Intuition and the Continuum." *Philosophia Mathematica* 8:259–273.

Benacerraf, Paul and Hilary Putnam, eds. 1964. *Philosophy of Mathematics: Selected Readings*. Englewood Cliffs, New Jersey: Prentice-Hall.

Bonola, Roberto. 1955. *Non-Euclidean Geometry: A Critical and Historical Study of its Development*. New York: Dover.

Borel, Armand. 2001. *Essays in the History of Lie Groups and Algebraic Groups.* Providence, Rhode Island: American Mathematical Society.

Bott, Raoul. 1988. "On Induced Representations." *Proceedings of Symposia in Pure Mathematics* 48:1–13.

Bourbaki, Nicolas. 1950. "The Architecture of Mathematics." *American Mathematical Monthly* 57:211–232.

Boyer, Carl. B. and Uta C. Merzbach. 1991. *A History of Mathematics*, 2nd. ed. New York: Wiley.

Brading, Katherine. 2002. "Which Symmetry? Noether, Weyl, and Conservation of Electric Charge." *Studies in the Philosophy of Modern Physics* 33:3–22.

Brading, Katherine and Harvey R. Brown. 2003a. "Symmetries and Noether's Theorem." In Brading and Castellani 2003, 89–109.

Brading, Katherine and Elena Castellani, eds. 2003b. *Symmetries in Physics: Philosophical Reflections.* Cambridge: Cambridge University Press.

Brauer, Richard and Hermann Weyl. 1935. "Spinors in n Dimensions." *American Journal of Mathematics* 57:425–449.

Brewer, James. W. and Martha K. Smith, eds. 1981. *Emmy Noether: A Tribute to her Life and Work.* New York: Marcel Dekker.

Brouwer, L. E. J. 1975. *Brouwer Collected Works*, ed. A. Heyting. Amsterdam: North-Holland.

Byers, Nina. 1996. "The Life and Times of Emmy Noether." In Newman and Ypsilantis 1996, 945–964.

Cantor, Georg. 1980. *Gesammelte Abhandlungen mathematischen und philosophischen Inhalt.* Berlin: Springer.

Cartan, Élie. 1931. "Le parallélisme absolu et la théorie unitaire du champ," *Revue de Métaphysique et de Morale* 38:13–28 (also in Cartan 1952–1955, 3(2):1167–1185).

—————. 1952–1955. *Œuvres complètes.* Paris: Gauthier-Villars.

—————. 1966. *The Theory of Spinors.* New York: Dover.

—————. 1983. *Geometry of Riemann Spaces*, edited by R. Hermann, tr. J. Glazebrook. Brookline, Mass.: Mathematical Sciences Press.

—————. 1986. *On Manifolds with an Affine Connection and the Theory of General Relativity*, tr. A. Magnon and A. Ashtekar. Naples: Bibliopolis.

Cartan, Élie and Albert Einstein. 1979. *Élie Cartan—Albert Einstein: Letters on Absolute Parallelism, 1929–1932*, ed. R. Debever. Princeton: Princeton University Press.

Cassirer, Ernst. 1942-1996. *The Philosophy of Symbolic Forms*, tr. Ralph Mannheim. New Haven: Yale University Press.

—————. 1953. *Substance and Function and Einstein's Theory of Relativity*, tr. William Curtis Swabey and Marie Collins Swabey. New York : Dover.

Catt, Carrie Chapman. 1935. *In Memory of Emmy Noether, Visiting Professor of Mathematics, Bryn Mawr College, 1933–April 1935.* Bryn Mawr: The College.

Chandrasekharan, K., ed. 1986. *Hermann Weyl: 1885–1985; Centenary Lectures delivered by C. N. Yang, R. Penrose, and A. Borel at the Eidgenössische Technische Hochschule Zürich.* Berlin: Springer.

Chatterji, S. D., ed. 1995. *Proceedings of the International Congress of Mathematicians, Zürich, Switzerland 1994.* Basel: Birkhäuser.

Chern, Shiing-Shen. 1945. "On the curvatura integra in Riemannian manifold." *Annals of Mathematics* 46: 674684.

Chevalley, Claude and André Weil. 1957. "Hermann Weyl (1885–1955)." *L'Enseignement* 3(3):157–187; *WGA* 4:655–685.

Chikara, Saski, Sugiura Mitsuo, and Joseph W. Dauben, eds. 1994. *The Intersection of History and Mathematics.* Basel: Birkhäuser.

Coleman, Robert and Herbert Korté. 2001. "Hermann Weyl: Mathematician, Physicist, Philosopher." In Scholz 2001a, 161–388.

Corry, Leo. 1999. "Hilbert and Physics (1900–1915)," in Gray 1999b, 145–188.

———. 2004a. *Modern Algebra and the Rise of Mathematical Structures,* 2nd revised ed. Basel: Birkhäuser.

———. 2004b. *David Hilbert and the Axiomatization of Physics (1898–1918): From* Grundlagen der Geometrie *to* Grundlagen der Physik. Dordrecht: Kluwer.

———. 2006. "Axiomatics, Empiricism and *Anschauung* in Hilbert's Conception of Geometry: Between Arithmetic and General Relativity." In Ferreirós and Gray 2006, 133–156.

Courant, R. 1950. *Dirichlet's Principle, Conformal Mapping, and Minimal Surfaces.* Appendix by M. Schiffer. New York: Interscience.

Courant, R. and David Hilbert. 1989. *Methods of Mathematical Physics.* New York: Wiley.

Cunningham, E. 1910. "The Principle of Relativity in Electrodynamics and an Extension Thereof." *Proceedings of the London Mathematical Society* 8:77–98.

Dawson, John W. Jr. 1997. *Logical Dilemmas: The Life and Work of Kurt Gödel.* Wellesley, Mass.: A. K. Peters.

Dyson, Freeman. 1956. "Obituary of Prof. Hermann Weyl, For.Mem.R.S." *Nature* 177:457–458.

Dedekind, Richard. 1963. *Essays on the Theory of Numbers,* tr. Wooster Woodruff Beman. New York: Dover.

Deppert, W., ed. 1988. *Exact Sciences and Their Philosophical Foundations.* Frankfurt am Main: Peter Lang.

Diamond, H. G. 1982. "Elementary methods in the study of the distribution of prime numbers." *Bulletin of the American Mathematical Society* 7:553-589.

Dick, Auguste. 1981. *Emmy Noether 1882–1935,* tr. H. I. Blocher. Boston: Birkhäuser.

Dieudonné, Jean. 1976. "Weyl, Hermann." In *Dictionary of Scientific Biography,* ed. C. G. Gillispie, 14:281-285.

Du Bois-Reymond, Emil. 1907. *Über die Grenzen des Naturerkennens. Die sieben Welträtsel.* Leipzig: Veit.

————. 1974. *Vorträge über Philosophie und Gesellschaft,* ed. Siegfried Wollgast. Hamburg: Meiner.

Eckes, Christophe. 2011. "Groupes, invariants et géométries dans l'oeuvre de Weyl." Ph.D. diss, University of Lyon.

Edward, Harold M. "Kronecker on the Foundations of Mathematics." In Hintikka 1995, 45–52.

Ehlers, Jürgen. 1988. "Hermann Weyl's Contribution to the General Theory of Relativity." In Deppert 1988, 83–105.

Ehrenfest, Paul. 1918. "In What Way Does It Become Manifest in the Fundamental Laws of Physics that Space Has Three Dimensions?" *Proceedings of the Royal Academy of Amsterdam* 20:200–209.

————. 1920. "Welche Rolle spielt die Dreidimensionalität des Raumes in den Grundgesetzen der Physik?" *Annalen der Physik,* 366:440–446.

Einstein, Albert. 1987– . *The Collected Papers of Albert Einstein.* Princeton: Princeton University Press. [Note that the page references for the translations in the companion volumes for each volume of main text will be included in square brackets.]

Ellison, W. J. 1971. "Waring's Problem." *American Mathematical Monthly* 78:10–36.

Ewald, William, ed. 1996. *From Kant to Hilbert: A Source Book in the Foundations of Mathematics.* Oxford: Clarendon Press.

Feferman, Solomon. 1998. *In the Light of Logic.* New York: Oxford University Press.

————. 2000. "The significance of Weyl's *Das Kontinuum.*" In Hendricks 2000, 179–194.

Feist, R. 2002. "Weyl's Appropriation of Husserl's and Poincaré's Thoughts." *Synthèse* 132:273–301.

Ferrierós, J. and J. J. Gray, eds. 2006. *The Architecture of Modern Mathematics: Essays in History and Philosophy.* Oxford: Oxford University Press.

Fogel, D. Brandon. 2008. "Epistemology of a Theory of Everything: Weyl, Einstein, and the Unification of Physics." Ph.D. diss., University of Notre Dame.

Forman, Paul. 1971. "Weimar Culture, Causality, and Quantum Theory 1918–1927: Adaptation by German Physicists and Mathematicians to a Hostile Intellectual Environment." *Historical Studies in the Physical Sciences* 3:1–116.

Frei, Gunther and Urs Stammbach. 1992. *Hermann Weyl und die Mathematik an der ETH Zürich, 1923–1930.* Basel: Birkhäuser.

Friedman, Michael. 1995. "Carnap and Weyl on the Foundations of Geometry and Relativity Theory." *Erkenntnis* 42:247–260.

————. 2001. *Foundations of Space-Time Theories: Relativistic Physics and Philosophy of Science*. Princeton: Princeton University Press.

Galilei, Galileo. 1957. *Discoveries and Opinions of Galileo*, tr. Stillman Drake. New York: Doubleday.

————. 1974. *Two New Sciences*, tr. Stillman Drake. Madison, Wisc.: University of Wisconsin Press.

Gavroglu, Kostas, Jean Christianidis, and Efthymios Nicolaidis, eds. 1994. *Trends in the Historiography of Science*. Dordrecht: Kluwer Academic.

Gavroglu, Kostas and Jürgen Renn, eds. 2007. *Positioning the History of Science*. Dordrecht: Springer.

Gel'fond, A. O. 1960. *Transcendental and Algebraic Numbers*, tr. Leo F. Boron. New York: Dover.

Gentzen, Gerhard. 1969. *The Collected Works of Gerhard Gentzen*, ed. M. E. Szabo. Amsterdam: North-Holland.

Giere, Ronald N. and Alan W. Richardson, eds. 1996. *Origins of Logical Empiricism. Minnesota Studies in the Philosophy of Science, volume XVI.* Minneapolis: University of Minnesota Press.

Goldfeld, Dorian. 2003. "The Elementary Proof of the Prime Number Theorem: An Historical Perspective." *Number Theory: New York Seminar*: 179192.

Gouvêa, Fernando Q. 1997. p-*adic Numbers: An Introduction*. 2nd ed. Berlin: Springer.

Gowers, Timothy, June Barrow-Green, and Imre Leader, eds. 2008. *The Princeton Companion to Mathematics*. Princeton: Princeton University Press.

Grattan-Guinness, Ivor, ed. 2005. *Landmark Writings in Western Mathematics, 1640–1940*. Amsterdam: Elsevier.

Gray, Jeremy. 1999a. "Geometry — Formalisms and Intuitions." In Gray 1999b, 58–83.

————, ed. 1999b. *The Symbolic Universe: Geometry and Physics 1890–1930*. Oxford: Oxford University Press.

————. 2000. *The Hilbert Challenge*. Oxford: Oxford University Press.

————. 2005. "Felix Klein's Erlangen Program, 'Comparative Considerations of Recent Geometrical Researches' (1872)," in Grattan-Guiness 2005b, 544–552.

————. 2008. *Plato's Ghost: The Modernist Transformation in Mathematics*. Princeton: Princeton University Press.

Hasse, Helmut. 1954. *Higher Algebra*, tr. Theodore J. Benac. New York: F. Ungar PUblishing Co.

Hawkins, Thomas. 1999. "Weyl and the Topology of Continuous Groups." In James 1999, 169–198.

————. 2000. *Emergence of the Theory of Lie Groups. An Essay in the History of Mathematics 1869–1926*. Berlin: Springer.

Helmholtz, Hermann von. 1883. *Wissenschaftliche Abhandlungen*. Leipzig: Johann Ambrosius Barth.

————. 1968. *Über Geometrie.* Darmstadt: Wissenschaftliche Buchgesellschaft.

————. 1971. *Selected Writings of Hermann von Helmholtz,* ed. Russell Kahl. Middletown, CT: Wesleyan University Press.

————. 1977. *Epistemological Writings,* translated by Malcolm F. Lowe, ed. Robert S. Cohen and Yehuda Elkana. Dordrecht: D. Reidel.

Hendricks, Vincent F., Stigandur Pedersen, and Klaus F. Jørgensen, eds. 2000. *Proof Theory: History and Philosophical Significance.* Dordrecht: Kluwer.

Hesseling, Dennis. 2003. *Gnomes in the Fog. The Reception of Brouwer's Intuitionism in the 1920s.* Basel: Birkhäuser.

Hilbert, David. 1932–1935. *Gesammelte Abhandlungen.* Berlin: J. Springer.

————. 1953. *Grundzüge einer allgemeinen Theorie der linearen Integralgleichungen.* New York: Chelsea Publishing.

————. 1971. *Foundations of Geometry,* tr. Leo Unger, rev. ed. Paul Bernays. La Salle, Il.: Open Court.

————. 1998. *The Theory of Algebraic Number Fields,* tr. Franz Lemmermeyer and Norbert Schappacher. Berlin: Springer-Verlag.

Hilbert, David and W. Ackermann. 1950. *Principles of Mathematical Logic.* New York: Chelsea.

Hilbert, David and Paul Bernays. 1968–1970. *Grundlagen der Mathematik.* Berlin: Springer.

Hilbert, David and Stephen Cohn-Vossen. 1999. *Geometry and the Imagination,* tr. P. Nemenyi. New York: Chelsea AMS.

Hintikka, Jaakko, ed. 1995. *From Dedekind to Gödel: Essays on the Development of the Foundations of Mathematics.* Dordrecht: Kluwer Academic.

Its, A. R. 2003. "The Riemann-Hilbert Problem and Integrable Systems." *Notices of the American Mathematical Society* 50:1389–1400.

James, I. M., ed. 1999. *History of Topology.* Amsterdam: Elsevier.

Kasner, Edward and James Newman. 1940. *Mathematics and the Imagination.* New York: Simon & Schuster. Reprinted 2001. Mineola, NY: Dover.

Kastrup, H. A. 2008. "On the Advancements of Conformal Transformations and their Associated Symmetries in Geometry and Theoretical Physics." *Annalen der Physik* 17:631–690.

Kleene, Stephen Cole. 1967. *Mathematical Logic.* New York: Wiley.

Klein, Felix. 1921–1923. *Gesammelte mathematische Abhandlungen.* Berlin: Springer.

————. 1939. *Elementary Mathematics from an Advanced Standpoint,* tr. E. R. Hedrick and C. A. Noble. New York: Dover.

————. 1956. *Lectures on the Icosahedron and the Solution of Equations of the Fifth Degree,* tr. George Gavin. New York: Dover.

————. 1979. *Development of Mathematics in the Nineteenth Century,* tr. M. Ackerman. Brookline, Mass.: Math Sci Press.

Klein, Jacob. 1985. *Lectures and Essays*, ed. Robert B. Williamson and Elliott Zuckerman. Annapolis, Md.: St. John's College Press.

————. 1992. *Greek Mathematical Thought and the Origin of Algebra*, tr. Eva Brann. New York: Dover.

Kleiner, Israel. 2007. *A History of Abstract Algebra*. Boston: Birkhäuser.

Kosmann-Schwarzbach, Yvette. 2004. *Les théorèmes de Noether: invariance et lois de conservations au XXe siècle*. Palaisseau: Éditions de l'École polytechnique.

Lämmerzahl, Claus and Alfredo Macias. 1993. "On the Dimensionality of Space-Time." *Journal of Mathematical Physics* 34:4540–4553.

Laugwitz, Detlef. 1958. "Über eine Vermutung von Hermann Weyl zum Raumproblem." *Archiv der Mathematik* 9:128–133.

Lefschetz, Solomon. 1956. *Topology*. New York: Chelsea Publishing.

Leupold, R. 1960. "Die Grundlagenforschung bei Hermann Weyl." Ph.D. diss., Universität Mainz.

Lützen, J., ed. 2006. *The Interaction between Mathematics, Physics and Philosophy from 1850 to 1940*. Dordrecht: Springer.

Mackey, George. 1988. "Hermnann Weyl and the application of group theory to quantum mechanics.". In Deppert 1988, 131–160.

Mancosu, Paolo. 1998. *From Brouwer to Hilbert. The Debate on the Foundations of Mathematics in the 1920s*. Oxford: Oxford University Press.

————. 2011. *The Adventure of Reason: Interplay Between Philosophy of Mathematics and Mathematical Logic, 1900–1940*. Oxford: Oxford University Press.

Martin, Christopher A. 2003. "On Continuous Symmetries and the Foundations of Modern Physics." In Brading and Castellani 2003, 29–60.

McClarty, Colin. 2006. "Emmy Noether's 'Set Theoretic' Topology: From Dedekind to the Rise of Functors." In Ferreirós and Gray 2006, 187–208.

Mielke, Eckehard W. and Friedrich W. Hehl. 1988. "Die Entwicklung der Eichtheorien: Marginalien zu deren Wissenschaftsgeschichte." In Deppert 1988, 191–230.

Millennial Conference on Number Theory. 2002. *Number Theory for the Millennium*. Natick, Mass.: A. K. Peters.

Minkowski, Hermann. 1968. *Geometrie der Zahlen*. New York: Johnson Reprint.

Narkiewicz, Władisław. 1988. "Hermann Weyl and the Theory of Numbers." In Deppert 1988, 51–60.

Nelson, William and Mairi Sakellariadou. 2009. "SpaceTime Dimensionality from Brane Collisions." *Physics Letters B* 674:210-212.

Newman, H. and T. Ypsilantis, eds. 1996. *History of Original Ideas and Basic Discoveries in Particle Physics*. New York: Plenum.

Newman, M. H. A. 1957. "Hermann Weyl, 1885–1955." *Biographical Memoirs of Fellows of the Royal Society London* 3:305–328.

O'Raifeartaigh, Lochlainn, ed. 1997. *The Dawning of Gauge Theory.* Princeton: Princeton University Press.

O'Raifeartaigh, Lochlainn and Norbert Straumann. 2000. "Gauge Theory: Historical Origins and Some Modern Developments." *Reviews of Modern Physics* 72:1–23.

Pais, Abraham. 1982. *'Subtle is the Lord . . . ': The Science and the Life of Albert Einstein.* Oxford: Oxford University Press.

———. 1986. *Inward Bound.* Oxford: Oxford University Press.

Peckhaus, Volker. 1990. *Hilbertprogramm und Kritische Philosophie: Das Göttinger Modell interdisziplinärer Zusammenarbeit zwischen Mathematik und Philosophie.* Göttingen: Vandenhoeck und Ruprecht.

Penrose, Roger. 1986. "Hermann Weyl, Space-Time, and Conformal Geometry." In Chandrasekharan 1986, 23–52.

Pesic, Peter. 1997. "Secrets, Symbols, and Systems: Parallels Between Cryptanalysis and Algebra, 1580-1700." *Isis* 88:674-692.

———. 2002. *Seeing Double: Shared Identities in Physics, Philosophy, and Literature.* Cambridge, Mass.: MIT Press.

———. 2003. *Abel's Proof: An Essay on the Sources and Meaning of Mathematical Unsolvability.* Cambridge, Mass.: MIT Press.

———, ed. 2007 *Beyond Geometry: Classic Papers from Riemann to Einstein.* Mineola, NY: Dover.

———. 2012. "Helmholtz, Riemann, and the Sirens: Sound, Color, and the Problem of Space." (unpublished essay)

Pesic, Peter and Stephen P. Boughn. 2003. "The Weyl-Cartan Theorem and the Naturalness of General Relativity," *European Journal of Physics* 24:261–266.

Pontrjagin, L. S. 1946. *Topological Groups,* tr. Emma Lehmer. Princeton: Princeton University Press.

Pound, R. V. and G. A. Rebka Jr. 1959. "Gravitational Red-Shift in Nuclear Resonance." *Physical Review Letters* 3:439-441.

Reale, Giovanni. 1990. *A History of Ancient Philosophy: Plato and Aristotle.* Albany: State University of New York Press.

Redhead, Michael. 2003. "The Interpretation of Gauge Symmetry." In Brading and Castellani 2003, 124–139.

Reich, Karin. 1992. "Levi-Civitasche Parallelverschiebung, affiner zusammenhang, Übertragungsprinzip: 1916/17–1922/23." *Archive for History of Exact Science* 44:77–105.

Reid, Constance. 1996. *Hilbert.* New York: Copernicus.

Ria, D. 2005. *L'unità fisico-matematica nel pensiero epistemologico di Hermann Weyl.* Lecce: Congedo Editore.

Riemann, Bernhard. 1919. *Über die Hypothesen, welche der Geometrie zu Grunde ligen,* ed. Hermann Weyl. Berlin: Springer (Reprint, 1923).

————. 1990. *Gesammelte mathematische Werke, wissenschaftlicher Nach-lass, und Nachträge*, ed. Raghavan Narasimhan after the edition by Heinrich Weber and Richard Dedekind. Berlin: Springer-Verlag.

Roquette, Peter. 2005. *The Brauer-Hasse-Noether Theorem in Historical Perspective*. Berlin: Springer.

————. 2008. "Emmy Noether and Hermann Weyl." In Tent 2006, 285–326.

Rowe, David E. 1985. "Felix Klein's 'Erlanger Antrittsrede': A Transcription with English Translation and Commentary." *Historia Mathematics* 12:123–141.

————. 1992. "Klein, Lie, and the 'Erlanger Programm'." In Boi, Flament, and Salaskis 1992, 45–54.

————. 1994. "The Philosophical Views of Klein and Hilbert." In Chikara et al. 1994, 187–202.

————. 1999. "The Göttingen Response to General Relativity and Emmy Noether's Theorem." In Gray 1999b, 189–223.

Rupke, Nicolaas, ed. 2002. *Göttingen and the Development of the Natural Sciences*. Göttingen: Wallstein Verlag.

Ryckman, Thomas A. 1996. "Einstein *Agonists*: Weyl and Reichenbach on Geometry and the General Theory of Relativity." In Giere and Richardson 1996, 165–209.

————. 2003. "The Philosophical Roots of the Gauge Principle: Weyl and Transcendental Phenomenological Idealism." In Brading and Castellani 2003, 61–88.

————. 2007. *The Reign of Relativity: Philosophy in Physics 1915–1925*. New York: Oxford University Press.

Sauer, Tilmann. 1999. "The Relativity of Discovery." *Archive for History of Exact Sciences* 53:529–575.

Scheibe, Erhard. 1988. "Hermann Weyl and the Nature of Spacetime." In Deppert 1988, 61–82.

Schilpp, Paul Arthur, ed. 1951. *The Philosophy of Bertrand Russell*, 3d ed. New York: Tudor.

Schlick, Moritz. 1963. *Space and Time in Contemporary Physics: An Introduction to the Theory of Relativity and Gravitation*, tr. H. L. Brose. New York: Dover.

Scholz, Erhard. 1980. *Geschichte des Mannigfaltigkeitsbegriff von Riemann bis Poincaré*. Basel: Birkhäuser.

————. 1994. "Hermann Weyl's Contributions to Geometry, 1917–1923." In Chikara et al. 1994, 203–230.

————. 1995. "Hermann Weyl's 'Purely Infinitesimal Geometry'." In Chatterji 1995, 1592–1603.

————. 1999a. "Weyl and the Theory of Connections." In Gray 1999b, 260–284.

————. 1999b. "The Concept of Manifold, 1850–1950." In James 1999, 25–64.

————. 2000. "Hermann Weyl on the Concept of Continuum." In Hendricks et al. 2000, 195–217.

————, ed. 2001a. *Hermann Weyl's Raum-Zeit-Materie and a General Introduction to His Scientific Work*. Basel: Birkhäuser.

————. 2001b. "Weyls Infinitesimalgeometrie, 1917–1925." In Scholz 2001a, 48–104.

————. 2004. "Hermann Weyl's Analysis of the 'Problem of Space' and the Origin of Gauge Structures." *Science in Context* 17:165–197.

————. 2005. "Philosophy as a Cultural Resource and Medium of Reflection for Hermann Weyl." *Revue de synthèse* 126:331–351.

————. 2006a. "Practice-Related Symbolic Realism in H. Weyl's Mature View of Mathematical Knowledge." In Ferreirós and Gray 2006, 291–310.

————. 2006b. "Introducing Groups into Quantum Theory (1926–1930)." *Historia Mathematica* 33:440–490.

————. 2006c. "The Changing Concept of Matter in H. Weyls Thought, 1918–1930." In Lützen 2006, 281–305.

————. 2009. "A. Einstein and H. Weyl: Intertwining Paths and Mutual Inuences." In Alunni et al. 2009, 215–230.

————. 2011a. "H. Weyl's and E. Cartan's Proposals for Infinitesimal Geometry in the Early 1920s." *Boletim da Sociedada Portuguesa de Matemática* Número Especial A. da Mira Fernandes, 225–245.

————. 2011b. "Mathematische Physik bei Hermann Weyl — zwischen 'Hegelscher Physik und 'symbolischer Konstruktion der Wirklichkeit." In Schlote and Schneider 2011, 183–212.

Schlote, K.-H., M. Schneider, eds. 2011. *Mathematics Meets Physics: A Contribution to Their Interaction in the 19th and the First Half of the 20th Century*. Frankfurt: Harri Deutsch Verlag.

Schwinger, Julian. 1988. "Hermann Weyl and Quantum Kinematics." In Deppert 1988, 107–129.

Shapiro, Stewarrt, ed. 2005. *The Oxford Handbook of Philosophy of Mathematics and Logic*. Oxford: Oxford University Press.

Shenitzer, Abe and John Stillwell, eds. 2002. *Mathematical Evolutions*. Washington, D.C.: Mathematical Association of America.

Sieroka, Norman. 2007. "Weyl's 'Agens Theory' of Matter and the Zurich Fichte." *Studies in the History and Philosophy of Science* 38:84–107.

————. 2009. "Husserlian and Fichtean Leanings: Weyl on Logicism, Intuitionism, and Formalism." *Philosophia Scientiae* 13:85–96.

————. 2010. *Umgebunden: symbolischer Konstruktivismus im Anschluss an Hermann Weyl und Fritz Medicus*. Zürich: Chronos.

————. 2012. "Hermann Weyl und Fritz Medicus: Die Zürcher Fichte-Interpretation in Mathematik und Physik um 1920." *Fichte-Studien* 36:129–143.

Sigurdsson, Skúli. 1991. "Hermann Weyl, Mathematics and Physics, 1900–1927." Ph.D. diss., Harvard University.

―――――. 1994. "Unification, Geometry and Ambivalence: Hilbert, Weyl and the Göttingen Community." In Gavroglu et al. 1994, 355–367.

―――――. 1996. "Physics, Life, and Contingency: Born, Schrödinger, and Weyl in Exile." In Ash and Söllner 1996, 48–70.

―――――. 2001. "Journeys in Spacetime." In Scholz 2001c, 15–47.

―――――. 2007. "On the Road." In Gavroglu and Renn 2007, 149–157.

Slodowy, Peter. 1999. "The Early Development of the Representation Theory of Semisimple Lie Groups: A. Hurwitz, I. Schur, H. Weyl." *Jahresbericht der Deutschen Matehematiker-Vereinigung* 101:97–115.

Speiser, David. 1988. "*Gruppentheorie und Quantenmechanik*: The Book and its Position in Weyl's Work." In Deppert 1988, 161–189.

Straumann, Norman. 2001. "Ursprünge der Eichtheorien." In Scholz 2001c, 138–160.

Suppes, Patrick, ed. 1973. *Space, Time, and Geometry*. Dordrecht: D. Reidel.

Tavel, M. A. 1971. "Noether's Theorem." *Transport Theory and Statistical Physics* 1:183–207.

Tegmark, Max. 1997. "On the dimensionality of spacetime." *Classical and Quantum Gravity* 14:L69-L75.

Tent, Katrin, ed. 2008. *Groups and Analysis: The Legacy of Hermann Weyl*. Cambridge: Cambridge University Press.

Tieszen, Richard. 2005. *Phenomenology, Logic, and the Philosophy of Mathematics*. Cambridge: Cambridge University Press.

Toader, Iulian D. 2011. "Objectivity Sans Intelligibility: Hermann Weyl's Symbolic Constructivism." Ph.D. diss., University of Notre Dame.

Tobies, Renate. 1981. *Felix Klein*. Leipzig: Teubner.

―――――. 2002. "The Development of Göttingen into the Prussian Centre of Mathematics and the Exact Sciences." In Ruppert 2002, 116–142.

Tonietti, Tino. 1988. "Four Letters of E. Husserl to H. Weyl and their Context." In Deppert 1988, 343–384.

Van Atten, Mark. 2007. *Brouwer Meets Husserl: On the Phenomenology of Choice Sequences*. Dordrecht: Springer.

Van Atten, Mark, Dirk van Dalen, and Richard Tieszen. 2002. "The Phenomenology and Mathematics of the Intuitive Continuum." *Philosophia Mathematica* 10:203–226.

Van Dalen, Dirk. 1984. "Four Letters from Edmund Husserl to Hermann Weyl." *Husserl Studies* 1:1–12.

―――――. 1995. "Hermann Weyl's Intuitionistic Mathematics." *Bulletin of Symbolic Logic* 1:145–169.

―――――. 1999, 2005. *Mystic, Geometer, and Intuitionist. The Life of L. E. J. Brouwer*. Oxford: Oxford University Press.

————. 2000. "Brouwer and Weyl on proof theory and philosophy of mathematics." In Hendricks et al. 2000, 117–152.

————. 2006. "The Genesis of Mathematical Objects, following Weyl and Brouwer." *La Nuova Critica* 48:5–18.

Vaughan, R. C. and Trevor Wooley. 2002. "Waring's Problem: A Survey". In Millennial Conference 2002, 201–340.

Veblen, Oswald. 1931. *Analysis Situs*. New York: American Mathematical Society.

Veblen, Oscar and Alfred North Whitehead. 1932. *The Foundations of Differential Geometry*. Cambridge: Cambridge University Press.

Vizgin, Vladimir. 1994. *Unified Field Theories in the First Third of the 20th Century*, tr. J. B. Barbour. Basel: Birkhäuser.

Waerden, B. L. van der. 1970. *Algebra*, tr. Fred Blum and John R. Schulenberger. New York: Ungar.

————. 1975. "On the Sources of My Book, *Moderne Algebra*." *Historia Mathematica* 2:31–40.

————. 1985. *A History of Algebra: From al-Khwārizmī to Emmy Noether*. Berlin: Springer-Verlag.

Wells, Raymond O., ed. 1988. *The Mathematical Heritage of Hermann Weyl: Proceedings of the Symposium on the Mathematical Heritage of Hermann Weyl, Held at Duke University, Durham, North Carolina, May 12–16, 1987*. Providence, RI: American Mathematical Society.

Weyl, Hermann. 1913. *Die Idee der Riemannschen Fläche*. Leipzig: Teubner. (Translation, Weyl 2009c)

————. 1918a. *Das Kontinuum. Kritische Untersuchungen über die Grundlagen der Analysis*. Leipzig: Veit. (Translation, Weyl 1994)

————. 1918b. *Raum-Zeit-Materie*. Berlin: Springer. (Translation, Weyl 1952a)

————. 1918c. "Gravitation und Elektrizität." *Sitzungsberichte der Königlich Preussischen Akademie der Wissenschaften zu Berlin* 465–480; *WGA* 2:29–42. (Translations in Lorentz et al. 1923, 200–216; O'Raifeartaigh 1997, 24–37.)

————. 1918d. "Reine Infinitesimalgeometrie." *Mathematische Zeitschrift* 2:384–411; *WGA* 2:1–28.

————. 1919a. *Erläuterungen in seiner Herausgabe von Riemann:* Hypothesen, welche der Geometrie zugrunde liegen. Berlin: Springer 1919. (Included in Riemann 1990, 740–769.)

————. 1919b. *Raum Zeit Materie*. 3rd ed. Berlin: Springer.

————. 1920a. "Elektrizität und Gravitation." *Physikalische Zeitschrift* 21:649–650; *WGA* 2:141–142.

————. 1920b. "Das Verhältnis der kausalen zur statistischen Betrachtungsweise in der Physik." *Schweizer Medizinische Wochenschrift*; *WGA* 2:113–122.

————. 1921a. "Electricity and Gravitation." *Nature* 106:800–802; *WGA* 2:260–262.

————. 1921b. "Uber die neue Grundlagenkrise der Mathematik." *Mathematische Zeitschrift* 10:39–79; *WGA* 2:143–180. (Translation in Mancosu 1998, 86–121)

————. 1923. *Mathematische Analyse des Raumproblems. Vorlesungen gehalten in Barcelona und Madrid.* Berlin: Springer. (Reprint, 1963)

————. 1927. *Philosophie der Mathematik und Naturwissenschaft, Handbuch der Philosophie, Abt. 2A.* München: Oldenbourg. (Translation in Weyl 1949, 2nd ed. 1966; new edition 2009b).

————. 1928. *Gruppentheorie und Quantenmechanik.* Leipzig: Hirzel. (Translation in Weyl 1950)

————. 1929a. "Gravitation and the Electron." *Proceedings of the National Academy of Sciences* 15:323–334; *WGA* 3:217–228.

————. 1929b. "Elektron und Gravitation." *Zeitschrift für Physik* 56:350–352; *WGA* 3:245–267.

————. 1931. *Die Stufen des Unendlichen.* Jena: Gustav Fischer.

————. 1935. "Emmy Noether 1882–1935." *Scripta mathematica* 3:201–220; *WGA* 3:425–444.

————. 1939a. *The Classical Groups, Their Invariants and Representations.* 2nd ed. 1946. Princeton: University Press.

————. 1939b. "Invariants." *Duke Mathematical Journal* 5:489–502; *WGA* 3:670–683.

————. 1940. "The Mathematical Way of Thinking," *Science* 92:437–446; *WGA* 3:710–718 (which prints a garbled text).

————. 1944a. "Obituary: David Hilbert (1862–1943)." *Obituary Notices of Fellows of the Royal Society* 4:547–553; *WGA* 4:121–129.

————. 1944b. "David Hilbert and his Mathematical Work." *Bulletin of the American Mathematical Society* 50:612–654; *WGA* 4:130–172.

————. 1946a. "Mathematics and Logic: A brief survey serving as a preface to a review of *The Philosophy of Bertrand Russell*," *American Mathematical Monthly* 53:2–13; *WGA* 4:268–279.

————. 1946b. "Review: *The Philosophy of Bertrand Russell*," *American Mathematical Monthly,* 53:208–214; *WGA* 4:599–608.

————. 1949a. *Philosophy of Mathematics and Natural Science,* tr. Olaf Helmer. Princeton: Princeton University Press. (2nd ed., 1966; new edition 2009b)

————. 1949b. "Relativity Theory as a Stimulus in Mathematical Research." *Proceedings of the American Philosophical Society* 93:35–541; *WGA* 4:394–400.

————. 1950. *The Theory of Groups and Quantum Mechanics,* tr. H. P. Robertson. New York: Dover.

————. 1951. "A Half-Century of Mathematics." *American Mathematical Monthly* 58:523–553; *WGA* 4:464–494.

————. 1952a. *Space-Time-Matter.* 4th German ed., tr. Henry L. Brose. New York: Dover.

————. 1952b. *Symmetry.* Princeton: Princeton University Press. (Reprint, 1980).

————. 1953a. "Über den Symbolismus der Mathematik und mathematischen Physik." *Studium generale* 6:219–228; *WGA* 4:527–536.

————. 1953b. "Universities and Science in Germany." *The Mathematics Student* (Madras, India) 21:1–26; *WGA* 4:537–562.

————. 1955. *Selecta.* Basel: Birkhäuser.

————. 1968. *Gesammelte Abhandlungen,* ed. K. Chandrasekharan. Berlin: Springer. Cited as *WGA.*

————. 1977. *Mathematische Analyse des Raumproblems/Was ist Materie?* Darmstadt: Wissenschaftliche Buchgesellschaft.

————. 1985. "Axiomatic Versus Constructive Procedures in Mathematics," ed. Tito Tonietti. *Mathematical Intelligencer* 7:10–17, 38.

————. 1988. *Riemanns geometrische Ideen, ihre Auswirkung und ihre Verknüpfung mit der Gruppentheorie (1925),* ed. K. Chandrasekharan. Berlin: Springer-Verlag.

————. 1994. *The Continuum,* tr. S. Pollard and T. Bole. New York: Dover.

————. 1995. "Topology and Abstract Algebra as Two Roads of Mathematical Comprehension," tr. Abe Shenitzer. *American Mathematical Monthly* 102:453–460, 646–651.

————. 1997. *The Classical Groups: Their Invariants and Representations.* Princeton: Princeton University Press.

————. 2009a. *Mind and Nature: Selected Writings on Philosophy, Mathematics, and Physics,* ed. Peter Pesic. Princeton: Princeton University Press.

————. 2009b. *Philosophy of Mathematics and Natural Science,* tr. Olaf Helmer. Princeton: Princeton University Press. (2nd ed., 1966; new edition with an introduction by Frank Wilczek)

————. 2009c. *The Concept of a Riemann Surface,* tr. Gerald R. Maclane. 3rd ed. Mineola, NY: Dover.

Wheeler, John A. 1988. "Hermann Weyl and the Unity of Knowledge." In Deppert 1988, 469–503. (An adapted version of this essay had appeared in 1986 in *American Scientist* 74:366–375.)

Whitrow, G. J. 1955. "Why Physical Space Has Three Dimensions." *British Journal for the Philosophy of Science* 6:13– 31.

Wittgenstein, Ludwig. 1975. *Philosophical Remarks,* ed. Rush Rhees. Oxford: Blackwell.

Wolfson, Harry Austryn. 1976. *The Philosophy of the Kalam.* Cambridge, Mass.: Harvard University Press.

Woodward, William R. 1978. "From Association to Gestalt: The Fate of Hermann Lotze's Theory of Spatial Perception, 1846–1920." *Isis* 69:572–582.

Wu, T. T. and Yang, C. N. 1975. "Concept of Non-Integrable Phase Factors and Global Formulation of Gauge Fields." *Physical Review D* 12:3845–3857.

Yaglom, I. M. 1988. *Felix Klein and Sophus Lie: Evolution of the Idea of Symmetry in the Nineteenth Century*. Boston: Birkhäuser.

Yang, Chen Ning. 1986. "Hermann Weyl's Contribution to Physics." In Chandrasekharan 1986, 7–21.

Zisman, M. 1999. "Fibre Bundles, Fibre Maps." In James 1999, 605–629.

Acknowledgments

We thank the following for their permission to reprint:

Die Stufe des Unendlichen, Jena: Gustav Fischer Verlag, 1931, translated and published with the permission of Annemarie Weyl Carr, Peter Weyl, and Thomas Weyl.

"Topology and Abstract Algebra as Two Roads of Mathematical Comprehension" (1932), as translated by Abe Shenitzer in *American Mathematical Monthly*, 102:453–460, 646–651 (1995). Reprinted with permission of The Mathematical Association of America.

"Mathematics and Logic," *American Mathematical Monthly*, 53:2–13 (1946). Reprinted with permission of The Mathematical Association of America.

"A Half-Century of Mathematics," *American Mathematical Monthly* 58:523–553 (1953). Reprinted with permission of The Mathematical Association of America.

"Emmy Noether," *Scripta mathematica* 3:201–220 (1935), reprinted with the permission of Yeshiva University.

"The Mathematical Way of Thinking," *Science* 92:437–446 (1940), reprinted with the permission of the American Association for the Advancement of Science.

"David Hilbert (1862-1943)," *Obituary Notices of Fellows of the Royal Society* 4:547–553 (1944), reprinted with permission of the Royal Society.

"David Hilbert and His Mathematical Work," Bulletin of the *American Mathematical Society* 50:612–654 (1944), reprinted with the permission of the American Mathematical Society.

"Relativity Theory as a Stimulus in Mathematical Research," *Proceedings of the American Philosophical Society* 93:535-541(1949), reprinted with the permission of the American Philosophical Society

"Axiomatic versus constructive procedures in mathematics" (Hs 91a:27) and "Why is the World Four-Dimensional?" (Hs 91a:33), unpublished manuscripts published with the permission in ETH-Bibliothek, Archive und Nächlasse

Index

Mathematics–Bestsellers

HANDBOOK OF MATHEMATICAL FUNCTIONS: with Formulas, Graphs, and Mathematical Tables, Edited by Milton Abramowitz and Irene A. Stegun. A classic resource for working with special functions, standard trig, and exponential logarithmic definitions and extensions, it features 29 sets of tables, some to as high as 20 places. 1046pp. 8 x 10 1/2. 0-486-61272-4

ABSTRACT AND CONCRETE CATEGORIES: The Joy of Cats, Jiri Adamek, Horst Herrlich, and George E. Strecker. This up-to-date introductory treatment employs category theory to explore the theory of structures. Its unique approach stresses concrete categories and presents a systematic view of factorization structures. Numerous examples. 1990 edition, updated 2004. 528pp. 6 1/8 x 9 1/4. 0-486-46934-4

MATHEMATICS: Its Content, Methods and Meaning, A. D. Aleksandrov, A. N. Kolmogorov, and M. A. Lavrent'ev. Major survey offers comprehensive, coherent discussions of analytic geometry, algebra, differential equations, calculus of variations, functions of a complex variable, prime numbers, linear and non-Euclidean geometry, topology, functional analysis, more. 1963 edition. 1120pp. 5 3/8 x 8 1/2. 0-486-40916-3

INTRODUCTION TO VECTORS AND TENSORS: Second Edition–Two Volumes Bound as One, Ray M. Bowen and C.-C. Wang. Convenient single-volume compilation of two texts offers both introduction and in-depth survey. Geared toward engineering and science students rather than mathematicians, it focuses on physics and engineering applications. 1976 edition. 560pp. 6 1/2 x 9 1/4. 0-486-46914-X

AN INTRODUCTION TO ORTHOGONAL POLYNOMIALS, Theodore S. Chihara. Concise introduction covers general elementary theory, including the representation theorem and distribution functions, continued fractions and chain sequences, the recurrence formula, special functions, and some specific systems. 1978 edition. 272pp. 5 3/8 x 8 1/2. 0-486-47929-3

ADVANCED MATHEMATICS FOR ENGINEERS AND SCIENTISTS, Paul DuChateau. This primary text and supplemental reference focuses on linear algebra, calculus, and ordinary differential equations. Additional topics include partial differential equations and approximation methods. Includes solved problems. 1992 edition. 400pp. 7 1/2 x 9 1/4. 0-486-47930-7

PARTIAL DIFFERENTIAL EQUATIONS FOR SCIENTISTS AND ENGINEERS, Stanley J. Farlow. Practical text shows how to formulate and solve partial differential equations. Coverage of diffusion-type problems, hyperbolic-type problems, elliptic-type problems, numerical and approximate methods. Solution guide available upon request. 1982 edition. 414pp. 6 1/8 x 9 1/4. 0-486-67620-X

VARIATIONAL PRINCIPLES AND FREE-BOUNDARY PROBLEMS, Avner Friedman. Advanced graduate-level text examines variational methods in partial differential equations and illustrates their applications to free-boundary problems. Features detailed statements of standard theory of elliptic and parabolic operators. 1982 edition. 720pp. 6 1/8 x 9 1/4. 0-486-47853-X

LINEAR ANALYSIS AND REPRESENTATION THEORY, Steven A. Gaal. Unified treatment covers topics from the theory of operators and operator algebras on Hilbert spaces; integration and representation theory for topological groups; and the theory of Lie algebras, Lie groups, and transform groups. 1973 edition. 704pp. 6 1/8 x 9 1/4. 0-486-47851-3

Browse over 9,000 books at www.doverpublications.com

A SURVEY OF INDUSTRIAL MATHEMATICS, Charles R. MacCluer. Students learn how to solve problems they'll encounter in their professional lives with this concise single-volume treatment. It employs MATLAB and other strategies to explore typical industrial problems. 2000 edition. 384pp. 5 3/8 x 8 1/2. 0-486-47702-9

NUMBER SYSTEMS AND THE FOUNDATIONS OF ANALYSIS, Elliott Mendelson. Geared toward undergraduate and beginning graduate students, this study explores natural numbers, integers, rational numbers, real numbers, and complex numbers. Numerous exercises and appendixes supplement the text. 1973 edition. 368pp. 5 3/8 x 8 1/2. 0-486-45792-3

A FIRST LOOK AT NUMERICAL FUNCTIONAL ANALYSIS, W. W. Sawyer. Text by renowned educator shows how problems in numerical analysis lead to concepts of functional analysis. Topics include Banach and Hilbert spaces, contraction mappings, convergence, differentiation and integration, and Euclidean space. 1978 edition. 208pp. 5 3/8 x 8 1/2. 0-486-47882-3

FRACTALS, CHAOS, POWER LAWS: Minutes from an Infinite Paradise, Manfred Schroeder. A fascinating exploration of the connections between chaos theory, physics, biology, and mathematics, this book abounds in award-winning computer graphics, optical illusions, and games that clarify memorable insights into self-similarity. 1992 edition. 448pp. 6 1/8 x 9 1/4. 0-486-47204-3

SET THEORY AND THE CONTINUUM PROBLEM, Raymond M. Smullyan and Melvin Fitting. A lucid, elegant, and complete survey of set theory, this three-part treatment explores axiomatic set theory, the consistency of the continuum hypothesis, and forcing and independence results. 1996 edition. 336pp. 6 x 9. 0-486-47484-4

DYNAMICAL SYSTEMS, Shlomo Sternberg. A pioneer in the field of dynamical systems discusses one-dimensional dynamics, differential equations, random walks, iterated function systems, symbolic dynamics, and Markov chains. Supplementary materials include PowerPoint slides and MATLAB exercises. 2010 edition. 272pp. 6 1/8 x 9 1/4. 0-486-47705-3

ORDINARY DIFFERENTIAL EQUATIONS, Morris Tenenbaum and Harry Pollard. Skillfully organized introductory text examines origin of differential equations, then defines basic terms and outlines general solution of a differential equation. Explores integrating factors; dilution and accretion problems; Laplace Transforms; Newton's Interpolation Formulas, more. 818pp. 5 3/8 x 8 1/2. 0-486-64940-7

MATROID THEORY, D. J. A. Welsh. Text by a noted expert describes standard examples and investigation results, using elementary proofs to develop basic matroid properties before advancing to a more sophisticated treatment. Includes numerous exercises. 1976 edition. 448pp. 5 3/8 x 8 1/2. 0-486-47439-9

THE CONCEPT OF A RIEMANN SURFACE, Hermann Weyl. This classic on the general history of functions combines function theory and geometry, forming the basis of the modern approach to analysis, geometry, and topology. 1955 edition. 208pp. 5 3/8 x 8 1/2. 0-486-47004-0

THE LAPLACE TRANSFORM, David Vernon Widder. This volume focuses on the Laplace and Stieltjes transforms, offering a highly theoretical treatment. Topics include fundamental formulas, the moment problem, monotonic functions, and Tauberian theorems. 1941 edition. 416pp. 5 3/8 x 8 1/2. 0-486-47755-X

Browse over 9,000 books at www.doverpublications.com

Mathematics–History

THE WORKS OF ARCHIMEDES, Archimedes. Translated by Sir Thomas Heath. Complete works of ancient geometer feature such topics as the famous problems of the ratio of the areas of a cylinder and an inscribed sphere; the properties of conoids, spheroids, and spirals; more. 326pp. 5 3/8 x 8 1/2. 0-486-42084-1

THE HISTORICAL ROOTS OF ELEMENTARY MATHEMATICS, Lucas N. H. Bunt, Phillip S. Jones, and Jack D. Bedient. Exciting, hands-on approach to understanding fundamental underpinnings of modern arithmetic, algebra, geometry and number systems examines their origins in early Egyptian, Babylonian, and Greek sources. 336pp. 5 3/8 x 8 1/2. 0-486-25563-8

THE THIRTEEN BOOKS OF EUCLID'S ELEMENTS, Euclid. Contains complete English text of all 13 books of the Elements plus critical apparatus analyzing each definition, postulate, and proposition in great detail. Covers textual and linguistic matters; mathematical analyses of Euclid's ideas; classical, medieval, Renaissance and modern commentators; refutations, supports, extrapolations, reinterpretations and historical notes. 995 figures. Total of 1,425pp. All books 5 3/8 x 8 1/2.
Vol. I: 443pp. 0-486-60088-2
Vol. II: 464pp. 0-486-60089-0
Vol. III: 546pp. 0-486-60090-4

A HISTORY OF GREEK MATHEMATICS, Sir Thomas Heath. This authoritative two-volume set that covers the essentials of mathematics and features every landmark innovation and every important figure, including Euclid, Apollonius, and others. 5 3/8 x 8 1/2.
Vol. I: 461pp. 0-486-24073-8
Vol. II: 597pp. 0-486-24074-6

A MANUAL OF GREEK MATHEMATICS, Sir Thomas L. Heath. This concise but thorough history encompasses the enduring contributions of the ancient Greek mathematicians whose works form the basis of most modern mathematics. Discusses Pythagorean arithmetic, Plato, Euclid, more. 1931 edition. 576pp. 5 3/8 x 8 1/2.
0-486-43231-9

CHINESE MATHEMATICS IN THE THIRTEENTH CENTURY, Ulrich Libbrecht. An exploration of the 13th-century mathematician Ch'in, this fascinating book combines what is known of the mathematician's life with a history of his only extant work, the Shu-shu chiu-chang. 1973 edition. 592pp. 5 3/8 x 8 1/2.
0-486-44619-0

PHILOSOPHY OF MATHEMATICS AND DEDUCTIVE STRUCTURE IN EUCLID'S ELEMENTS, Ian Mueller. This text provides an understanding of the classical Greek conception of mathematics as expressed in Euclid's Elements. It focuses on philosophical, foundational, and logical questions and features helpful appendixes. 400pp. 6 1/2 x 9 1/4. 0-486-45300-6

BEYOND GEOMETRY: Classic Papers from Riemann to Einstein, Edited with an Introduction and Notes by Peter Pesic. This is the only English-language collection of these 8 accessible essays. They trace seminal ideas about the foundations of geometry that led to Einstein's general theory of relativity. 224pp. 6 1/8 x 9 1/4. 0-486-45350-2

HISTORY OF MATHEMATICS, David E. Smith. Two-volume history – from Egyptian papyri and medieval maps to modern graphs and diagrams. Non-technical chronological survey with thousands of biographical notes, critical evaluations, and contemporary opinions on over 1,100 mathematicians. 5 3/8 x 8 1/2.
Vol. I: 618pp. 0-486-20429-4
Vol. II: 736pp. 0-486-20430-8

Mathematics–Logic and Problem Solving

PERPLEXING PUZZLES AND TANTALIZING TEASERS, Martin Gardner. Ninety-three riddles, mazes, illusions, tricky questions, word and picture puzzles, and other challenges offer hours of entertainment for youngsters. Filled with rib-tickling drawings. Solutions. 224pp. 5 3/8 x 8 1/2.   0-486-25637-5

MY BEST MATHEMATICAL AND LOGIC PUZZLES, Martin Gardner. The noted expert selects 70 of his favorite "short" puzzles. Includes The Returning Explorer, The Mutilated Chessboard, Scrambled Box Tops, and dozens more. Complete solutions included. 96pp. 5 3/8 x 8 1/2.   0-486-28152-3

THE LADY OR THE TIGER?: and Other Logic Puzzles, Raymond M. Smullyan. Created by a renowned puzzle master, these whimsically themed challenges involve paradoxes about probability, time, and change; metapuzzles; and self-referentiality. Nineteen chapters advance in difficulty from relatively simple to highly complex. 1982 edition. 240pp. 5 3/8 x 8 1/2.   0-486-47027-X

SATAN, CANTOR AND INFINITY: Mind-Boggling Puzzles, Raymond M. Smullyan. A renowned mathematician tells stories of knights and knaves in an entertaining look at the logical precepts behind infinity, probability, time, and change. Requires a strong background in mathematics. Complete solutions. 288pp. 5 3/8 x 8 1/2.

0-486-47036-9

THE RED BOOK OF MATHEMATICAL PROBLEMS, Kenneth S. Williams and Kenneth Hardy. Handy compilation of 100 practice problems, hints and solutions indispensable for students preparing for the William Lowell Putnam and other mathematical competitions. Preface to the First Edition. Sources. 1988 edition. 192pp. 5 3/8 x 8 1/2.   0-486-69415-1

KING ARTHUR IN SEARCH OF HIS DOG AND OTHER CURIOUS PUZZLES, Raymond M. Smullyan. This fanciful, original collection for readers of all ages features arithmetic puzzles, logic problems related to crime detection, and logic and arithmetic puzzles involving King Arthur and his Dogs of the Round Table. 160pp. 5 3/8 x 8 1/2.

0-486-47435-6

UNDECIDABLE THEORIES: Studies in Logic and the Foundation of Mathematics, Alfred Tarski in collaboration with Andrzej Mostowski and Raphael M. Robinson. This well-known book by the famed logician consists of three treatises: "A General Method in Proofs of Undecidability," "Undecidability and Essential Undecidability in Mathematics," and "Undecidability of the Elementary Theory of Groups." 1953 edition. 112pp. 5 3/8 x 8 1/2.   0-486-47703-7

LOGIC FOR MATHEMATICIANS, J. Barkley Rosser. Examination of essential topics and theorems assumes no background in logic. "Undoubtedly a major addition to the literature of mathematical logic." – Bulletin of the American Mathematical Society. 1978 edition. 592pp. 6 1/8 x 9 1/4.   0-486-46898-4

INTRODUCTION TO PROOF IN ABSTRACT MATHEMATICS, Andrew Wohlgemuth. This undergraduate text teaches students what constitutes an acceptable proof, and it develops their ability to do proofs of routine problems as well as those requiring creative insights. 1990 edition. 384pp. 6 1/2 x 9 1/4.   0-486-47854-8

FIRST COURSE IN MATHEMATICAL LOGIC, Patrick Suppes and Shirley Hill. Rigorous introduction is simple enough in presentation and context for wide range of students. Symbolizing sentences; logical inference; truth and validity; truth tables; terms, predicates, universal quantifiers; universal specification and laws of identity; more. 288pp. 5 3/8 x 8 1/2.   0-486-42259-3

Mathematics–Probability and Statistics

BASIC PROBABILITY THEORY, Robert B. Ash. This text emphasizes the probabilistic way of thinking, rather than measure-theoretic concepts. Geared toward advanced undergraduates and graduate students, it features solutions to some of the problems. 1970 edition. 352pp. 5 3/8 x 8 1/2. 0-486-46628-0

PRINCIPLES OF STATISTICS, M. G. Bulmer. Concise description of classical statistics, from basic dice probabilities to modern regression analysis. Equal stress on theory and applications. Moderate difficulty; only basic calculus required. Includes problems with answers. 252pp. 5 5/8 x 8 1/4. 0-486-63760-3

OUTLINE OF BASIC STATISTICS: Dictionary and Formulas, John E. Freund and Frank J. Williams. Handy guide includes a 70-page outline of essential statistical formulas covering grouped and ungrouped data, finite populations, probability, and more, plus over 1,000 clear, concise definitions of statistical terms. 1966 edition. 208pp. 5 3/8 x 8 1/2. 0-486-47769-X

GOOD THINKING: The Foundations of Probability and Its Applications, Irving J. Good. This in-depth treatment of probability theory by a famous British statistician explores Keynesian principles and surveys such topics as Bayesian rationality, corroboration, hypothesis testing, and mathematical tools for induction and simplicity. 1983 edition. 352pp. 5 3/8 x 8 1/2. 0-486-47438-0

INTRODUCTION TO PROBABILITY THEORY WITH CONTEMPORARY APPLICATIONS, Lester L. Helms. Extensive discussions and clear examples, written in plain language, expose students to the rules and methods of probability. Exercises foster problem-solving skills, and all problems feature step-by-step solutions. 1997 edition. 368pp. 6 1/2 x 9 1/4. 0-486-47418-6

CHANCE, LUCK, AND STATISTICS, Horace C. Levinson. In simple, non-technical language, this volume explores the fundamentals governing chance and applies them to sports, government, and business. "Clear and lively ... remarkably accurate." – *Scientific Monthly.* 384pp. 5 3/8 x 8 1/2. 0-486-41997-5

FIFTY CHALLENGING PROBLEMS IN PROBABILITY WITH SOLUTIONS, Frederick Mosteller. Remarkable puzzlers, graded in difficulty, illustrate elementary and advanced aspects of probability. These problems were selected for originality, general interest, or because they demonstrate valuable techniques. Also includes detailed solutions. 88pp. 5 3/8 x 8 1/2. 0-486-65355-2

EXPERIMENTAL STATISTICS, Mary Gibbons Natrella. A handbook for those seeking engineering information and quantitative data for designing, developing, constructing, and testing equipment. Covers the planning of experiments, the analyzing of extreme-value data; and more. 1966 edition. Index. Includes 52 figures and 76 tables. 560pp. 8 3/8 x 11. 0-486-43937-2

STOCHASTIC MODELING: Analysis and Simulation, Barry L. Nelson. Coherent introduction to techniques also offers a guide to the mathematical, numerical, and simulation tools of systems analysis. Includes formulation of models, analysis, and interpretation of results. 1995 edition. 336pp. 6 1/8 x 9 1/4. 0-486-47770-3

INTRODUCTION TO BIOSTATISTICS: Second Edition, Robert R. Sokal and F. James Rohlf. Suitable for undergraduates with a minimal background in mathematics, this introduction ranges from descriptive statistics to fundamental distributions and the testing of hypotheses. Includes numerous worked-out problems and examples. 1987 edition. 384pp. 6 1/8 x 9 1/4. 0-486-46961-1

Physics

THEORETICAL NUCLEAR PHYSICS, John M. Blatt and Victor F. Weisskopf. An uncommonly clear and cogent investigation and correlation of key aspects of theoretical nuclear physics by leading experts: the nucleus, nuclear forces, nuclear spectroscopy, two-, three- and four-body problems, nuclear reactions, beta-decay and nuclear shell structure. 896pp. 5 3/8 x 8 1/2.   0-486-66827-4

QUANTUM THEORY, David Bohm. This advanced undergraduate-level text presents the quantum theory in terms of qualitative and imaginative concepts, followed by specific applications worked out in mathematical detail. 655pp. 5 3/8 x 8 1/2.
0-486-65969-0

ATOMIC PHYSICS AND HUMAN KNOWLEDGE, Niels Bohr. Articles and speeches by the Nobel Prize–winning physicist, dating from 1934 to 1958, offer philosophical explorations of the relevance of atomic physics to many areas of human endeavor. 1961 edition. 112pp. 5 3/8 x 8 1/2.   0-486-47928-5

COSMOLOGY, Hermann Bondi. A co-developer of the steady-state theory explores his conception of the expanding universe. This historic book was among the first to present cosmology as a separate branch of physics. 1961 edition. 192pp. 5 3/8 x 8 1/2.
0-486-47483-6

LECTURES ON QUANTUM MECHANICS, Paul A. M. Dirac. Four concise, brilliant lectures on mathematical methods in quantum mechanics from Nobel Prize-winning quantum pioneer build on idea of visualizing quantum theory through the use of classical mechanics. 96pp. 5 3/8 x 8 1/2.   0-486-41713-1

THE PRINCIPLE OF RELATIVITY, Albert Einstein and Frances A. Davis. Eleven papers that forged the general and special theories of relativity include seven papers by Einstein, two by Lorentz, and one each by Minkowski and Weyl. 1923 edition. 240pp. 5 3/8 x 8 1/2.   0-486-60081-5

PHYSICS OF WAVES, William C. Elmore and Mark A. Heald. Ideal as a classroom text or for individual study, this unique one-volume overview of classical wave theory covers wave phenomena of acoustics, optics, electromagnetic radiations, and more. 477pp. 5 3/8 x 8 1/2.   0-486-64926-1

THERMODYNAMICS, Enrico Fermi. In this classic of modern science, the Nobel Laureate presents a clear treatment of systems, the First and Second Laws of Thermodynamics, entropy, thermodynamic potentials, and much more. Calculus required. 160pp. 5 3/8 x 8 1/2.   0-486-60361-X

QUANTUM THEORY OF MANY-PARTICLE SYSTEMS, Alexander L. Fetter and John Dirk Walecka. Self-contained treatment of nonrelativistic many-particle systems discusses both formalism and applications in terms of ground-state (zero-temperature) formalism, finite-temperature formalism, canonical transformations, and applications to physical systems. 1971 edition. 640pp. 5 3/8 x 8 1/2.   0-486-42827-3

QUANTUM MECHANICS AND PATH INTEGRALS: Emended Edition, Richard P. Feynman and Albert R. Hibbs. Emended by Daniel F. Styer. The Nobel Prize–winning physicist presents unique insights into his theory and its applications. Feynman starts with fundamentals and advances to the perturbation method, quantum electrodynamics, and statistical mechanics. 1965 edition, emended in 2005. 384pp. 6 1/8 x 9 1/4.   0-486-47722-3

Browse over 9,000 books at www.doverpublications.com

Physics

INTRODUCTION TO MODERN OPTICS, Grant R. Fowles. A complete basic undergraduate course in modern optics for students in physics, technology, and engineering. The first half deals with classical physical optics; the second, quantum nature of light. Solutions. 336pp. 5 3/8 x 8 1/2. 0-486-65957-7

THE QUANTUM THEORY OF RADIATION: Third Edition, W. Heitler. The first comprehensive treatment of quantum physics in any language, this classic introduction to basic theory remains highly recommended and widely used, both as a text and as a reference. 1954 edition. 464pp. 5 3/8 x 8 1/2. 0-486-64558-4

QUANTUM FIELD THEORY, Claude Itzykson and Jean-Bernard Zuber. This comprehensive text begins with the standard quantization of electrodynamics and perturbative renormalization, advancing to functional methods, relativistic bound states, broken symmetries, nonabelian gauge fields, and asymptotic behavior. 1980 edition. 752pp. 6 1/2 x 9 1/4. 0-486-44568-2

FOUNDATIONS OF POTENTIAL THERY, Oliver D. Kellogg. Introduction to fundamentals of potential functions covers the force of gravity, fields of force, potentials, harmonic functions, electric images and Green's function, sequences of harmonic functions, fundamental existence theorems, and much more. 400pp. 5 3/8 x 8 1/2.
0-486-60144-7

FUNDAMENTALS OF MATHEMATICAL PHYSICS, Edgar A. Kraut. Indispensable for students of modern physics, this text provides the necessary background in mathematics to study the concepts of electromagnetic theory and quantum mechanics. 1967 edition. 480pp. 6 1/2 x 9 1/4. 0-486-45809-1

GEOMETRY AND LIGHT: The Science of Invisibility, Ulf Leonhardt and Thomas Philbin. Suitable for advanced undergraduate and graduate students of engineering, physics, and mathematics and scientific researchers of all types, this is the first authoritative text on invisibility and the science behind it. More than 100 full-color illustrations, plus exercises with solutions. 2010 edition. 288pp. 7 x 9 1/4. 0-486-47693-6

QUANTUM MECHANICS: New Approaches to Selected Topics, Harry J. Lipkin. Acclaimed as "excellent" (*Nature*) and "very original and refreshing" (*Physics Today*), these studies examine the Mössbauer effect, many-body quantum mechanics, scattering theory, Feynman diagrams, and relativistic quantum mechanics. 1973 edition. 480pp. 5 3/8 x 8 1/2. 0-486-45893-8

THEORY OF HEAT, James Clerk Maxwell. This classic sets forth the fundamentals of thermodynamics and kinetic theory simply enough to be understood by beginners, yet with enough subtlety to appeal to more advanced readers, too. 352pp. 5 3/8 x 8 1/2. 0-486-41735-2

QUANTUM MECHANICS, Albert Messiah. Subjects include formalism and its interpretation, analysis of simple systems, symmetries and invariance, methods of approximation, elements of relativistic quantum mechanics, much more. "Strongly recommended." – *American Journal of Physics.* 1152pp. 5 3/8 x 8 1/2. 0-486-40924-4

RELATIVISTIC QUANTUM FIELDS, Charles Nash. This graduate-level text contains techniques for performing calculations in quantum field theory. It focuses chiefly on the dimensional method and the renormalization group methods. Additional topics include functional integration and differentiation. 1978 edition. 240pp. 5 3/8 x 8 1/2.
0-486-47752-5

Physics

MATHEMATICAL TOOLS FOR PHYSICS, James Nearing. Encouraging students' development of intuition, this original work begins with a review of basic mathematics and advances to infinite series, complex algebra, differential equations, Fourier series, and more. 2010 edition. 496pp. 6 1/8 x 9 1/4. 0-486-48212-X

TREATISE ON THERMODYNAMICS, Max Planck. Great classic, still one of the best introductions to thermodynamics. Fundamentals, first and second principles of thermodynamics, applications to special states of equilibrium, more. Numerous worked examples. 1917 edition. 297pp. 5 3/8 x 8. 0-486-66371-X

AN INTRODUCTION TO RELATIVISTIC QUANTUM FIELD THEORY, Silvan S. Schweber. Complete, systematic, and self-contained, this text introduces modern quantum field theory. "Combines thorough knowledge with a high degree of didactic ability and a delightful style." – *Mathematical Reviews.* 1961 edition. 928pp. 5 3/8 x 8 1/2. 0-486-44228-4

THE ELECTROMAGNETIC FIELD, Albert Shadowitz. Comprehensive undergraduate text covers basics of electric and magnetic fields, building up to electromagnetic theory. Related topics include relativity theory. Over 900 problems, some with solutions. 1975 edition. 768pp. 5 5/8 x 8 1/4. 0-486-65660-8

THE PRINCIPLES OF STATISTICAL MECHANICS, Richard C. Tolman. Definitive treatise offers a concise exposition of classical statistical mechanics and a thorough elucidation of quantum statistical mechanics, plus applications of statistical mechanics to thermodynamic behavior. 1930 edition. 704pp. 5 5/8 x 8 1/4. 0-486-63896-0

INTRODUCTION TO THE PHYSICS OF FLUIDS AND SOLIDS, James S. Trefil. This interesting, informative survey by a well-known science author ranges from classical physics and geophysical topics, from the rings of Saturn and the rotation of the galaxy to underground nuclear tests. 1975 edition. 320pp. 5 3/8 x 8 1/2. 0-486-47437-2

STATISTICAL PHYSICS, Gregory H. Wannier. Classic text combines thermodynamics, statistical mechanics, and kinetic theory in one unified presentation. Topics include equilibrium statistics of special systems, kinetic theory, transport coefficients, and fluctuations. Problems with solutions. 1966 edition. 532pp. 5 3/8 x 8 1/2. 0-486-65401-X

SPACE, TIME, MATTER, Hermann Weyl. Excellent introduction probes deeply into Euclidean space, Riemann's space, Einstein's general relativity, gravitational waves and energy, and laws of conservation. "A classic of physics." – *British Journal for Philosophy and Science.* 330pp. 5 3/8 x 8 1/2. 0-486-60267-2

RANDOM VIBRATIONS: Theory and Practice, Paul H. Wirsching, Thomas L. Paez and Keith Ortiz. Comprehensive text and reference covers topics in probability, statistics, and random processes, plus methods for analyzing and controlling random vibrations. Suitable for graduate students and mechanical, structural, and aerospace engineers. 1995 edition. 464pp. 5 3/8 x 8 1/2. 0-486-45015-5

PHYSICS OF SHOCK WAVES AND HIGH-TEMPERATURE HYDRO DYNAMIC PHENOMENA, Ya B. Zel'dovich and Yu P. Raizer. Physical, chemical processes in gases at high temperatures are focus of outstanding text, which combines material from gas dynamics, shock-wave theory, thermodynamics and statistical physics, other fields. 284 illustrations. 1966–1967 edition. 944pp. 6 1/8 x 9 1/4. 0-486-42002-7

Browse over 9,000 books at www.doverpublications.com

Astronomy

CHARIOTS FOR APOLLO: The NASA History of Manned Lunar Spacecraft to 1969, Courtney G. Brooks, James M. Grimwood, and Loyd S. Swenson, Jr. This illustrated history by a trio of experts is the definitive reference on the Apollo spacecraft and lunar modules. It traces the vehicles' design, development, and operation in space. More than 100 photographs and illustrations. 576pp. 6 3/4 x 9 1/4.   0-486-46756-2

EXPLORING THE MOON THROUGH BINOCULARS AND SMALL TELESCOPES, Ernest H. Cherrington, Jr. Informative, profusely illustrated guide to locating and identifying craters, rills, seas, mountains, other lunar features. Newly revised and updated with special section of new photos. Over 100 photos and diagrams. 240pp. 8 1/4 x 11.   0-486-24491-1

WHERE NO MAN HAS GONE BEFORE: A History of NASA's Apollo Lunar Expeditions, William David Compton. Introduction by Paul Dickson. This official NASA history traces behind-the-scenes conflicts and cooperation between scientists and engineers. The first half concerns preparations for the Moon landings, and the second half documents the flights that followed Apollo 11. 1989 edition. 432pp. 7 x 10.
                    0-486-47888-2

APOLLO EXPEDITIONS TO THE MOON: The NASA History, Edited by Edgar M. Cortright. Official NASA publication marks the 40th anniversary of the first lunar landing and features essays by project participants recalling engineering and administrative challenges. Accessible, jargon-free accounts, highlighted by numerous illustrations. 336pp. 8 3/8 x 10 7/8.   0-486-47175-6

ON MARS: Exploration of the Red Planet, 1958-1978--The NASA History, Edward Clinton Ezell and Linda Neuman Ezell. NASA's official history chronicles the start of our explorations of our planetary neighbor. It recounts cooperation among government, industry, and academia, and it features dozens of photos from Viking cameras. 560pp. 6 3/4 x 9 1/4.   0-486-46757-0

ARISTARCHUS OF SAMOS: The Ancient Copernicus, Sir Thomas Heath. Heath's history of astronomy ranges from Homer and Hesiod to Aristarchus and includes quotes from numerous thinkers, compilers, and scholasticists from Thales and Anaximander through Pythagoras, Plato, Aristotle, and Heraclides. 34 figures. 448pp. 5 3/8 x 8 1/2.
                    0-486-43886-4

AN INTRODUCTION TO CELESTIAL MECHANICS, Forest Ray Moulton. Classic text still unsurpassed in presentation of fundamental principles. Covers rectilinear motion, central forces, problems of two and three bodies, much more. Includes over 200 problems, some with answers. 437pp. 5 3/8 x 8 1/2.   0-486-64687-4

BEYOND THE ATMOSPHERE: Early Years of Space Science, Homer E. Newell. This exciting survey is the work of a top NASA administrator who chronicles technological advances, the relationship of space science to general science, and the space program's social, political, and economic contexts. 528pp. 6 3/4 x 9 1/4.
                    0-486-47464-X

STAR LORE: Myths, Legends, and Facts, William Tyler Olcott. Captivating retellings of the origins and histories of ancient star groups include Pegasus, Ursa Major, Pleiades, signs of the zodiac, and other constellations. "Classic." – Sky & Telescope. 58 illustrations. 544pp. 5 3/8 x 8 1/2.   0-486-43581-4

A COMPLETE MANUAL OF AMATEUR ASTRONOMY: Tools and Techniques for Astronomical Observations, P. Clay Sherrod with Thomas L. Koed. Concise, highly readable book discusses the selection, set-up, and maintenance of a telescope; amateur studies of the sun; lunar topography and occultations; and more. 124 figures. 26 halftones. 37 tables. 335pp. 6 1/2 x 9 1/4.   0-486-42820-6

Browse over 9,000 books at www.doverpublications.com